Sustainable Civil Infrastructures

AF789950

About this Series

Sustainable Infrastructure impacts our well-being and day-to-day lives. The infrastructures we are building today will shape our lives tomorrow. The complex and diverse nature of the impacts due to weather extremes on transportation and civil infrastructures can be seen in our roadways, bridges, and buildings. Extreme summer temperatures, droughts, flash floods, and rising numbers of freeze-thaw cycles pose challenges for civil infrastructure and can endanger public safety. We constantly hear how civil infrastructures need constant attention, preservation, and upgrading. Such improvements and developments would obviously benefit from our desired book series that provide sustainable engineering materials and designs. The economic impact is huge and much research has been conducted worldwide. The future holds many opportunities, not only for researchers in a given country, but also for the worldwide field engineers who apply and implement these technologies. We believe that no approach can succeed if it does not unite the efforts of various engineering disciplines from all over the world under one umbrella to offer a beacon of modern solutions to the global infrastructure. Experts from the various engineering disciplines around the globe will participate in this series, including: Geotechnical, Geological, Geoscience, Petroleum, Structural, Transportation, Bridge, Infrastructure, Energy, Architectural, Chemical and Materials, and other related Engineering disciplines.

More information about this series at http://www.springer.com/series/15140

Kun Zhang · Rongqiao Xu
Shih-Huang Chen
Editors

Testing and Characterization of Asphalt Materials and Pavement Structures

Proceedings of the 5th GeoChina International Conference 2018 – Civil Infrastructures Confronting Severe Weathers and Climate Changes: From Failure to Sustainability, held on July 23 to 25, 2018 in HangZhou, China

Springer

Editors
Kun Zhang
Washington State University
Pullman, WA, USA

Shih-Huang Chen
National Central University
Taoyuan, Taiwan

Rongqiao Xu
Zhejiang University
Hangzhou, China

ISSN 2366-3405 ISSN 2366-3413 (electronic)
Sustainable Civil Infrastructures
ISBN 978-3-319-95788-3 ISBN 978-3-319-95789-0 (eBook)
https://doi.org/10.1007/978-3-319-95789-0

Library of Congress Control Number: 2018948638

Printed on acid-free paper

This Springer imprint is published by the registered company Springer International Publishing AG part of Springer Nature
The registered company address is: Gewerbestrasse 11, 6330 Cham, Switzerland

Contents

Fuel Resistance Asphalt Binder: Mixing Procedure and Fuel Damage Resistance 1
Hashim Rizvi, Y. Mehta, D. Weis, C. Purdy and A. Ali

Evaluating the Thermal Characteristics of Rubberized Asphalt by Applying the Object-Based Approach 12
Shenming Wu, Yishuo Huang, Yu-Min Su and Yuan-Zhih Lin

Effect of Anti-stripping Agents on Asphalt Mix Performance Using a Mechanistic Approach 21
Shivani Rani, Rouzbeh Ghabchi, Syed Ashik Ali and Musharraf Zaman

Stiffening Thin Orthotropic Deck Structures with Thermoset Epoxy Asphalt for Improved Fatigue Resistance 32
Can Chen, Wolfgang O. Eisenhut, Kreisler Lau, Alex Ingram and John Bors

Changes in Chemical Fingerprints of Asphalt Binders Due to Aging 44
Shahriar Alam and Zahid Hossain

Asphalt Mixtures that Dissipates Energy—Comparison of Conventional and Newly Developed Mixtures 55
Josef Zak, Jan Suda, Ondrej Dasek and Petr Spacek

Study on Moisture Susceptibility of Foamed Warm Mix Asphalt 69
Yingbiao Wu, Qingyi Xiao, Jinjin Shi and Peiliang Zhang

Effect of Different Adding Methods of Nano-TiO_2 on Photocatalytic Degradation of Automobile Exhaust on Asphalt Mixture 79
Qingyi Xiao, Xiangyang Chen, Yun Yang and Jinguo Zhao

Evaluating the Effect of Solar Radiations on the Performance of Asphalt Concrete Pavements 90
Amal Abdelaziz and Chun-Hsing Ho

Effect of Overloaded Vehicles on Whole Life Cycle Cost of Flexible Pavements 104
Dawid Rys and Piotr Jaskula

Practical Pavement Design Approach Subjected to Seepage Conditions 118
Anand Tapase and M. S. Ranadive

Study on Shrinkage of Ordinary Concrete Under Different Temperatures and Humidity 128
Jinjin Shi, Shitao Liu, Tongxin Liang, Hongling Yu and Ying Zhang

Representing Collected Road Condition Data with Chernoff Faces for Evaluation of Pavement Conditions 136
Aioub H. Guha and Gabriel J. Assaf

Reflective Cracking in Composite Pavements—A Case Study 151
Rongyao Du

Optimization of Gravel Road Blading 163
Wynand J. vd M. Steyn

Author Index 171

Introduction

Asphalt materials have been widely used to construct flexible pavements. The mechanical and chemical characterizations of asphalt binders and mixtures and the analysis of pavement structures and distresses are of great importance to building sustainable flexible pavements. The research studies presented in this volume deal with the works made by pavement researchers and engineers to address the contemporary issues in pavement engineering, such as aging of asphalt binders, applications of rubberized asphalt and epoxy-modified asphalt, performance evaluations of warm mix asphalt, mechanical-based pavement structure analysis, etc. The papers presented in this volume include five studies on asphalt binders, four studies on asphalt mixtures, and six studies on pavement design and distress analysis. This volume is part of the proceedings of the 5th GeoChina International Conference on Civil Infrastructures Confronting Severe Weathers and Climate Changes: From Failure to Sustainability, HangZhou, China 2018.

Keywords: Pavement engineering • Characterization of asphalt mixtures
Modification of asphalt materials • Pavement structural analysis
Pavement distresses analysis

Fuel Resistance Asphalt Binder: Mixing Procedure and Fuel Damage Resistance

Hashim Rizvi[1], Y. Mehta[1(✉)], D. Weis[2], C. Purdy[2], and A. Ali[1]

[1] Department of Civil and Environmental Engineering, Center for Research and Education in Advanced Transportation Engineering Systems (CREATEs), Rowan University, Glassboro, NJ, USA
{rizvi, mehta, alia}@rowan.edu
[2] Department of Civil and Environmental Engineering, Rowan University, Glassboro, NJ, USA
weisd5@students.rowan.edu, purdyc51@rowan.edu

Abstract. Most of the biopolymers are made from animal waste. Animal waste is treated and converted in a form of glue generally known as Hide Glue or Bone Glue (BG), which is prepared from animal protein. This study focusses of developing fuel resistant asphalt using biopolymers i.e. protein based polymers. A procedure to develop fuel resistant asphalt for commercial use and its performance evaluation was conducted in this study. Mixing procedure of BG was evaluated and refined and multiple fuel resistance tests were performed to evaluate the impact of bone glue modification on resistance to fuel of neat binder. It was determined that BG-binder mixing equipment developed in earlier studies was appropriate. Optimum BG dosage to develop FRA was found to be higher than the BG dosage recommended in earlier studies to develop mechanistically better performing asphalt binder. Fuel solubility resistance test was used to select the optimum BG dosage. Fourier Transform Infrared Spectroscopy results proved that water was completely evaporated using the mixing procedure developed for the preparation of FRA. Fuel damage resistance test revealed that BG modification not only improved the intermediate to high temperature characteristics of the neat binder but outperformed the conventional polymer modified asphalt binder in resistance to fuel damage.

1 Introduction and Background

Asphalt cement is a cementitious material either naturally occurring or produced through petroleum distillation. Asphalt is a black, sticky, semisolid, and a highly viscous material at ambient temperature (Brown et al. 2009). The Asphalt binder is used in construction industry especially in road pavements. Like any other material, asphalt binder also has some limitations such as; rutting at high temperatures, brittle at lower temperatures, and soluble in hydrocarbon fuels. High and low temperature problems are controlled by improving the rut resistance and fatigue performance of the binder, respectively. However, resistance to hydrocarbon fuel solubility is not well explored. The asphalt binder chemistry i.e. carbon (>80%), hydrogen (>10%) and sulfur ($\sim$5%) makes it a viscous hydrocarbon i.e. soluble in all petroleum fuels. Therefore, asphalt

K. Zhang et al. (eds.), *Testing and Characterization of Asphalt Materials and Pavement Structures*, Sustainable Civil Infrastructures,
https://doi.org/10.1007/978-3-319-95789-0_1

industry uses hydrocarbon fuels to reduce the viscosity of asphalt for lower application temperatures. This phenomenon is termed as cutback asphalts.

The use of hydrocarbon fuel to develop cutback asphalts is surely an advantage; however, the solubility of asphalt binder in hydrocarbon fuel is detrimental to pavement sustainability especially where frequent oil spills occur such as: airport taxiways, aircraft hangers, aircraft gates, car parking lots, private property driveways etc. The hydrocarbon fuel damages the asphalt binder and in turn, binder loses its binding strength, which is critical for mixture stability. Resistance of asphalt binder to solubility in hydrocarbon fuel has been explored by few researchers (Shoenberger 1983; Corun et al. 2006; Rooijen et al. 2004). Till now, Fuel Resistant Asphalt (FRA) has been explored as a replacement to the conventional flexible pavement on airport pavements. A FRA binder was developed by a private firm in the name Sealoflex®SFB 5-JR-50. The specifications of the binder and development recipe has not been shared; however it is proven that they used synthetic polymers to develop this FRA (Corun et al. 2006).

This study focusses of developing FRA using biopolymers i.e. protein based polymers. Most of the biopolymers are made from animal waste (Rizvi et al. 2014, 2015). Animal waste is treated and converted in a form of glue generally known as Hide Glue or Bone Glue (BG) (Rizvi et al. 2014) (Fig. 1a), which is prepared from animal protein. The BG has been used to study the effects of BG-modification on performance characteristics of asphalt binder and it was determined that BG modification improved the mechanical and rheological properties of asphalt binder (Rizvi et al. 2014, 2015). However, Rizvi et al. did not explore the potential of BG-modified binder as FRA. The concept of using biopolymer to develop FRA is based on the fact that protein is not soluble in hydrocarbon fuels (Cecil and Louis 1970); therefore, BG-modified asphalt binder was selected to evaluate its resistance to fuel damage.

Fig. 1. **a** Bone glue **b** BGW mixture heating and **c** heating mantle **d** oil bath

It has been a known fact that protein is soluble in aqueous solutions (i.e. based on water) (Nick et al. 2004). Therefore, the water was used to achieve the homogeneous mixing of asphalt binder and BG. Rizvi et al. (2014) developed a mixing procedure; however various equipment, temperature, and mixing times may be explored to determine the optimum mixing conditions for BG mixing. Moreover, fuel damage resistance of the BG modified asphalt binder was also explored to determine the resistance to fuel capability of BG modified binder. In summary, this study explored optimum mixing conditions, equipment, and fuel damage resistance capabilities of BG-modified binder.

2 Objectives

The objective of this study is to optimize the mixing procedure of BG and explore the fuel damage resistance capabilities of BG modified binder.

3 Materials and Testing

Two frequently used binders i.e. PG 64-22 (Un-modified Binder (Neat)), and PG 76-22 (Polymer Modified Binder (PMB)), were selected to be used for this study. The BG was purchased from local vendor in finely ground form. The fuel used for this study was automobile 87 octane gasoline (AKI rating 87-Octane).

The mixing procedure was adopted from the Rizvi et al. (2014) and heating equipment, temperature and mixing time was altered to achieve the optimum mixing conditions and BG dosage. Procedure developed by Rizvi et al. and alterations to the procedure are provided in Table 1. The lower BG dosage i.e. 2.5% was eliminated and 30% was added to explore the optimum resistance to fuel damage. Intermediate dosage was also eliminated i.e. 7.5% because Rizvi et al. proved that 10% was the optimum dosage to develop BG modified asphalt binder. Mixing equipment was also changed to determine the effect of equipment. Heating mantle was selected because heating mantle is simple and easy to use in comparison to oil bath i.e. more expensive, cumbersome and energy consuming (as it take some time to heat up the oil). Three Mixing temperatures were used: one for heating mantle i.e. 180 °C; two for the oil bath i.e. 140 and 175 °C. The objective of changing the temperature was to reduce the mixing time. Lastly, two mixing times were selected to determine the optimum time. Rizvi et al. selected 70 min for 10% and reported that BG dosage and mixing time are directly proportional to each other. In this study, consistent time was used for all BG dosages and the temperature was increased and mixing time was reduced to study the effect. Rest of the mixing procedure parameters were not altered.

Optimum BG dosage was based upon the maximum resistance to fuel solubility resistance. A fuel solubility resistance test was developed in this study to determine the optimum BG dosage for fuel solubility resistance by calculating the percentage of binder lost due to fuel exposure in 3, 6 and 24 h (Table 2). The details of the test are provided as follows:

- Neat and BG modified binders were heated, poured into sealable pans, and allowed to cool to room temperature (25 ± 5 °C). Fuel, equal in mass of binder, was added to the binder and the pan was sealed, weighed, and placed in an environmental chamber at 25 ± 5 °C. The pans were taken out of the environmental chamber after 3, 6, and 24 h and the liquid (i.e. fuel-binder mixture) was poured. Shaking and jolting were avoided to minimize variability. A flame-ignition was applied (i.e. not

Table 1. Mixing procedure comparison for BG and asphalt binder

Mixing procedure to make BG-modified binder	
Rizvi et al. (2014)	Variations in this study
BG submerged in water for 15 min	–
Water quantity was two times the weight of BG	–
BG-water mixture was heated for 10 min (continuous stir to prevent burning) (Fig. 1b)	–
BG poured in preheated asphalt binder at 150 °C	–
BG dosage (2.5, 5, 7.5, 10 and 20%)	5, 10, 15, 20, 30%
Mixing conducted using oil bath	Heating mantle (Fig. 1c) and oil bath (Fig. 1d)
Mixing temperature 130 °C	180 °C for heating mantle 145 and 175 °C for oil bath
Mixing times 70 min	40 and 60 min

Table 2. Types of test and respective parameters

Description	Parameters
Fuel solubility resistance	Dissolved percentage of the binder in fuel in 3, 6 and 24 h
FTIR spectroscopy	Water peaks
G*-master curve	Rheological index (R); cross-over frequency (ω_c)

touching the surface of the binder) to let the trapped fuel burn. It was observed that a very weak flame lasted from two to three seconds on the top of the mixture in pan. Two consecutive flames-ignitions were introduced in intervals enough to let the mixture cool down to room temperature (i.e. approximately two minutes). For each trial, 40 g of binder was used for FSR test because anything greater than 40 g of mixture combined with the fuel would not fit in the testing tins. The testing temperature was 25 ± 5 °C because it was easy to manage the binder storage and testing. Furthermore, higher temperatures might be hazardous due to flammable materials. The pan containing the binder was weighed after the fuel-binder mixture was spilled to analyze the binder's resistance to fuel. The fuel solubility resistance results are measured in terms of percent binder mass loss and are calculated by the following equation:

$$P_L = \frac{B_i - B_f}{B_i} * 100 \quad (1)$$

where

P_L = Binder loss, %
B_i = Initial binder mass, g
B_f = Final binder mass after testing, g

The optimum mixing conditions were based on the complete or maximum evaporation of water from the mixture; therefore, Fourier Transform Infrared (FTIR) spectroscopy was used to trace the presence of water in mixture (Table 2). FTIR test was conducted using Perkin Elmer Frontier FTIR Spectrometer. The FTIR spectrum of asphalt composite shows Carbon-Hydrogen stretching and bending frequencies characteristic of a hydrocarbon (Rizvi et al. 2014).

The damage caused by fuel on asphalt was evaluated based on: (1) Comparison of parameters determined using the G*-master curve of un-soaked and fuel soaked binders developed using time-temperature superposition methodology; and (2) Comparison of Fourier Transform Infrared (FTIR) spectroscopy of hydrocarbon fuel peaks between un-soaked and fuel soaked binder. The concept of fuel damage resistance test was adopted from Rooijen et al. (2004). However, details and quantities of material and procedure is provided as follows:

- A 40 g sample of binder was poured into a pan followed by a 40 g sample of fuel, completely soaking the binder in fuel. The fuel quantity was determined by trial and error method so that fuel would be sufficient to soak the binder completely. The pans containing binder-fuel mixture were left unsealed in an environmentally controlled chamber to let the fuel evaporate at room temperature for five days. Dynamic Shear Rheometer (DSR) was used to conducted frequency sweep tests at six frequencies and five temperatures i.e. (0.1, 0.5, 1, 5, 10, 25 Hz & 4, 10, 20, 35, 50 °C). G*-master curves of fuel damaged and undamaged binders were developed and curve shape, rheological index (R) and crossover frequency (ω_c) were discussed to evaluate the effects of fuel damage on performance of asphalt binders (Table 2).

4 Results and Discussion

4.1 Mixing Procedure

Mixing procedure of BG and asphalt binder results proved that due to high stiffness at 20 and 30% dosages, the 20% BG dosage was not workable. Furthermore, at 30% dosage, binder was not mixable. Therefore, no further mixing trials or testing was performed using more than 15% of BG dosage and three dosages were experimented further i.e. 5, 10 and 15%.

4.2 Fuel Solubility Resistance Test

The results of FSR test are provided in Fig. 2. It can be observed that BG15% showed least binder loss among all dosages of BG proving that 15% is the optimum dosage to develop FRA. BG modification reduced the binder loss due to solubility in hydrocarbon fuel up to 41.6% after exposure of 24 h. The percent binder loss of BG15% was less than neat binder by 20.5, 21 and 41.6% after 3, 6 and 24 h of fuel exposure, respectively. These results proved that 15% BG dosage approximately doubled the FSR capabilities of the neat binder after 24 h of fuel exposure. As the time of soaking increases, the fuel diffuses into the binder and breaks carbon chains causing it to dissolve. This process of diffusion may take hours to complete. It is evident from FSR test results that initially the diffusion might not be complete; therefore the results show that after 3 and 6 h, the FSR enhancement is approximately 20%. On the other hand, when diffusion process was given time, as observed in case of neat binder, where approximately 90% of the binder was lost after 24 h, the BG modification showed its real impact and resulted in a two-fold improvement in FSR capability of neat binders.

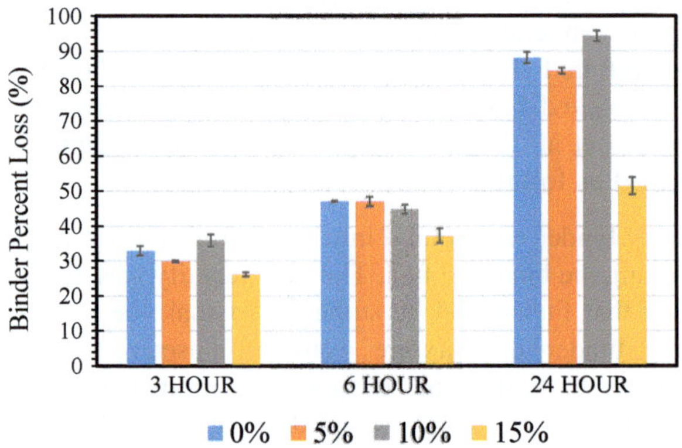

Fig. 2. Fuel solubility resistance test results at different dosages of bone glue modified binder

Figure 2 also showed that 5 and 10% BG dosages did not significantly impact the FSR capabilities of the neat binder. In order to determine the statistical significance of the mean differences of the results of three replicates, T-test between neat, 5 and 10% results were conducted and it was found that all *p*-values were higher than 0.05, proving that the mean difference between neat and 5%, and neat and 10% is not statistically significant. However, the t-test between neat and 15% BGB showed *p*-value 0.001, proving that FSR of 15% bone glue modified binder was statistically greater than neat binder. In summary, the results of FSR test proved that minimum dosage to observe fuel resistance capabilities is 15%; therefore only 15% BG dosage was selected for further testing.

4.3 Fourier Transform Infrared (FTIR) Spectroscopy

As mentioned above, 15% BG dosage was selected for further testing; therefore, FTIR spectroscopy (Fig. 2) was conducted on neat and 15% BGB binders to determine optimum mixing conditions for BG and binder. Generally, IR spectrum covers a range of wavelengths from 650 to 4000 cm^{-1}. Peaks in IR spectrum show the stretching and bending frequencies of different chemical compounds in specific range such as water–O–H can be observed in a range of 3100–3700 cm^{-1} (Rizvi et al. 2014) and IR stretch from 1500 to 1750 cm^{-1} shows C=O bonding that covers anhydrides, ketones, amides and. ester. It is essential to note that excess of the C=O bond shows oxidative aging of asphalt binder (Li et al. 2016). The thermo-oxidative aging decreases the content of small molecules and increases the content of large asphalt molecules; thus, leading to an increase in the average molecular weight and size of the asphalt binders (Li et al. 2016; Dehouche et al. 2012).

IR spectrum results are provided in Fig. 3. The figure is divided in four sections A, B, C and D representing combined, equipment, temperature and time effects on composition of binder, respectively. The water–O–H (3100–3700 cm^{-1}) and C=O (1500–1700 cm^{-1}) frequency stretch in IR spectrum (highlighted in A) were used to compare different mixing procedures. Combined spectrum (A) results proved that all the procedures successfully evaporated up to 99% of water because significantly bigger peaks cannot be observed in the water–O–H bonding range except B-HM-40 mix, which represents mix prepared in heating mantle for 40 min. However, significant peaks were observed in C=O bonding range showing oxidative aging of binder due to mixing procedure (Fig. 2a; C=O Bonding Zone).

Spectrum B shows that the heating mantle could not completely evaporate the water at 185 °C in 70 min (blue) as a water peak was found in case of B-HM-40 mix, as opposed to the B140-40 (mustard),which was prepared using an oil bath. The reason for this seems to be limited heating zone i.e. only bottom of the beaker was being heated; whereas the rest of the beaker was exposed to the room temperature. On the other hand, in the oil bath the beaker was fully submerged in preheated oil providing direct heating not only to the bottom but also to the sides. This result proved that using an oil bath is more appropriate a wet mixing procedure.

It can be observed in spectrum C that B175-40 mix, prepared in oil bath at 175 °C for 40 min, showed a stronger peak in the C=O bonding range in comparison to B140-40 mix, which was prepared in oil bath at 140 °C for 40 min. IR spectrum analysis proved the occurrence of extensive oxidative aging at 175 °C, represented by a green line in the spectrum. However, mix prepared at 140 °C (red) showed no peak in comparison to Neat binder (black). It was determined that evaporation of total water content remained consistent at both temperatures; therefore, to achieve the goal of complete water evaporation while simultaneously avoiding unnecessary oxidative aging, a 140 °C mixing temperature is optimum for BGB. Effect of mixing time i.e. spectrum D shows results of neat (black) and two mixes B140-40 (red) and B140-60 (blue), prepared at 140 °C for 40 and 60 min, respectively. Significant peaks observed for B140-60 in C=O bonding range proved that extended mixing time induced oxidative aging; whereas, B140-40 did not show any peak in C=O bonding range. This proved that even a 20 min difference could be significant for oxidative aging. It is evident from these results that BG should be mixed in oil bath at 140 °C for 40 min.

Fuel Damage Resistance Evaluation

The master curves were developed using the frequency sweep test data and shown in Fig. 4. The average percentage difference in stiffness over the broad range of frequencies between a pair of unsoaked and fuel-soaked neat, PMB and BG15% binders was found to be 97, 88 and 76%, respectively. Therefore, the stiffness reduction from fuel exposure was 30 and 20% lower in BG15% binder in comparison to neat and PMB binders, respectively. The BG modification enhanced the fuel resistance of neat binder and showed even better performance than a conventional PMB modified with SBS.

The drop in stiffness due to soaking in fuel is more significant at lower frequency/higher temperature zone than at higher frequency/low temperature zone

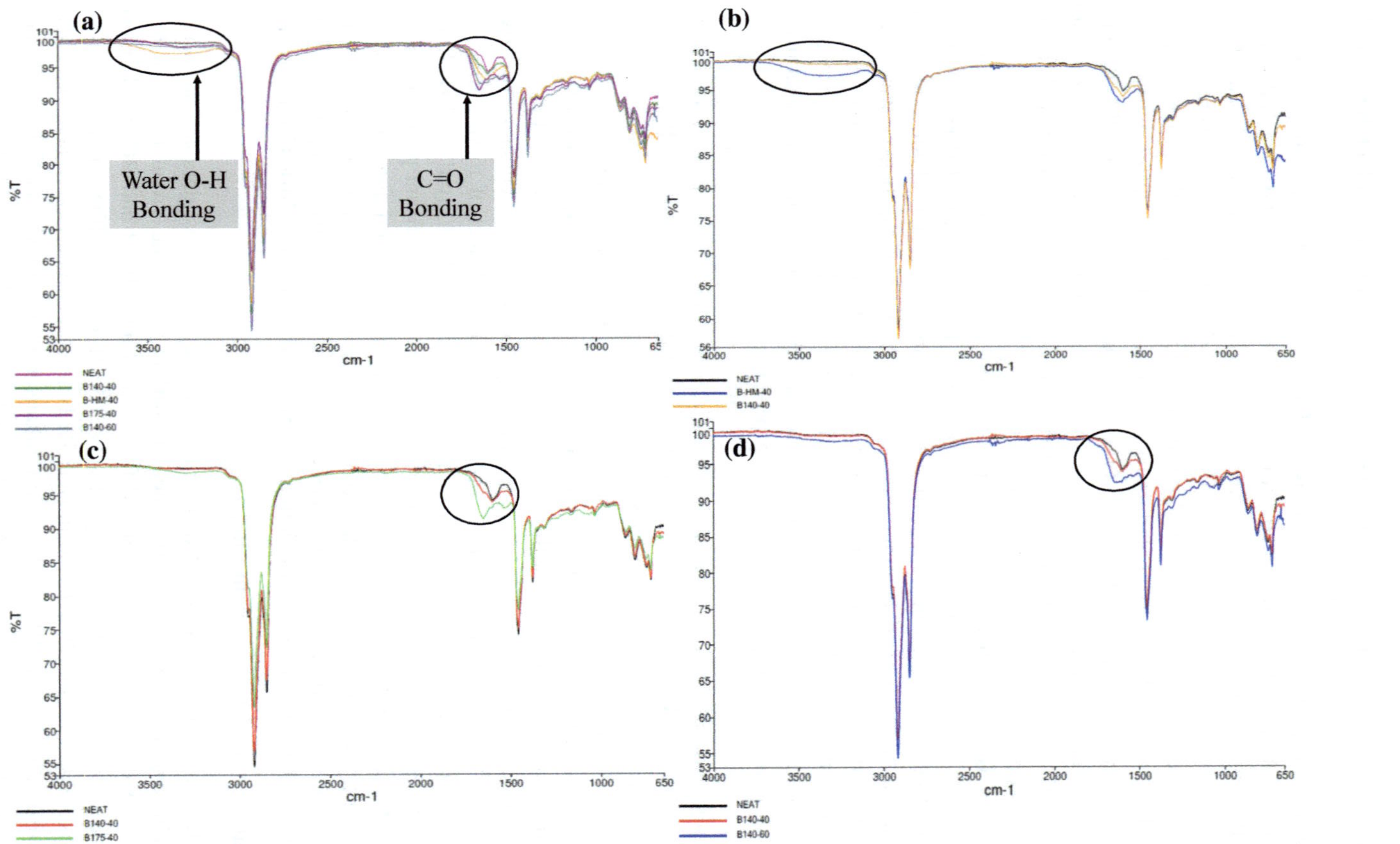

Note: B refers to 15% BGB; HM represents Heating Mantle

Fig. 3. FTIR spectroscopy of various BGB mixes

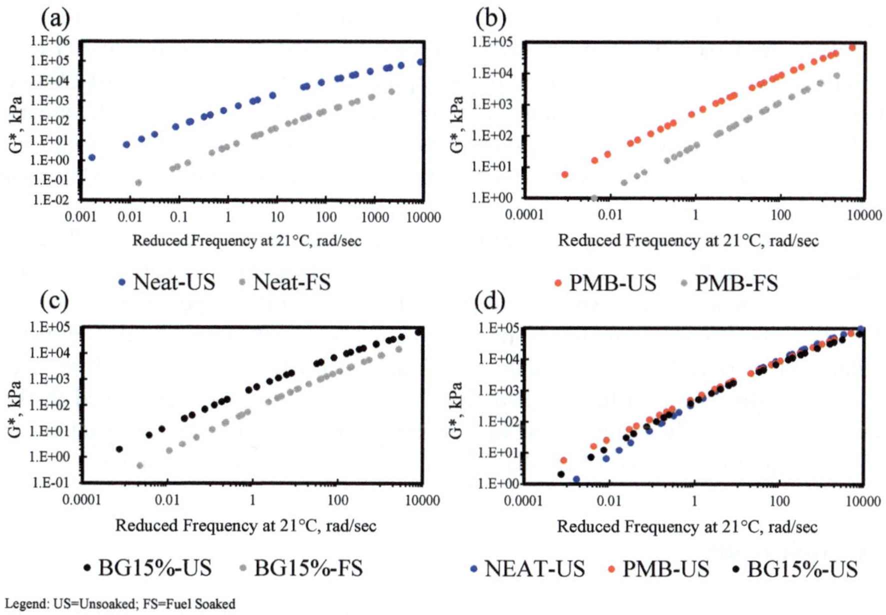

Fig. 4. G* master curve comparison of all binders

(Fig. 4a–c). This was due to effect of high temperature on viscosity of the binder providing conducive environment for fuel damage. On the contrary, higher frequency/low temperature zone is where binder behaves as a solid therefore less reduction in stiffness was observed due to fuel exposure in all binders. The reduction in stiffness of the BG15% fuel-soaked binder was found to 11 and 5% less than the fuel-soaked neat and PMB binders, respectively. Generally, intermediate to high temperature zone is more critical in terms of fuel damage due to softening of binder and it was observed that BG modification offered 30% higher resistance to fuel damage. Nevertheless, bio-polymer showed better fuel resistance in comparison to the conventional synthetic polymers in both high and low frequency zones.

Although PMB showed that fuel damage resistance can also be achieved through synthetic polymer modification, BG modification surpassed the conventional polymer modification impact on fuel damage resistance. A possible reason for this is that the protein based bio-polymer chains are more fuel resistant in nature. The insolubility of protein-fibrous chains in petroleum fuels provided better resistance to fuel damage in comparison to conventional synthetic polymer chains.

The master curves of un-soaked binders showed that BG modification showed approximately 70% increase in stiffness of neat binder at lower frequency and high temperatures, hence improving rutting performance. The stiffness of BG modified binders were found to be approximately 30% lower than the neat and PMB binders at higher frequencies and lower temperatures. Therefore, master curve data proved that the BG modification did not show improvement in binder properties of neat binder at lower temperatures. Rheological index (R) and cross-over frequency (ω_c) were also calculated from fuel soaked master curves of the respective binders using the Christensen-Anderson (CA) model (Christensen and Anderson 1992). The BG modification enhanced the R value of neat binder approximately ten times i.e. 40–390 kPa. This shows significant improvement in rut resistance on the neat binder. Please note that BG15% showed higher R-value than that of the PMB by approximately four times, which proved that BG15% performed better than the conventional PMB in rut resistance.

The Cross-Over frequency (ω_c) represents hardness parameter. BG modification exhibited more stiffening effect in comparison to neat binder i.e. neat and BG15% cross-over frequency values were found to be 30 and 17 rad/s, respectively. Therefore, fuel soaked BG15% was found to be harder than the fuel soaked neat binder. However, calculated cross-over frequency value of fuel soaked PMB was 12 rad/s. This results shows that performance of fuel soaked BG15% is better than the fuel soaked PMB at lower temperature. This shows that fuel soaking has larger negative impact on low temperature performance of PMB than on 15% bone glue modified binder.

5 Conclusions

This study was focused on developing fuel resistant asphalt for commercial use. Animal waste, Bone Glue, was selected to be used for this purpose and multiple fuel resistance tests were performed to evaluate the impact of bone glue modification on resistance to fuel of neat binder. In conclusion;

1. Optimum bone glue dosage in asphalt binder was found to be 15% to develop fuel resistant asphalt binder.
2. Mixing procedure developed by Rizvi et al. was found to be viable except change in mixing time and temperature. It was determined that an increment of 10 °C in mixing temperature reduced the mixing time to approximately half of the recommended time of 70 min.
3. Oil bath was found to be appropriate mixing equipment for BG and binder mixing process. Heating mantle was easy to use but the results proved that BG and binder mixture was not homogeneous and it could also not evaporate the water from the mix.
4. Fuel solubility resistance was significantly improved by BG modification.
5. It was also proven that BG modification outperformed the conventional synthetic polymer modification by showing higher resistance to fuel damage. It can be concluded that BG modification can transform neat binder into a FRA.
6. It was determined that low temperature characteristics of PMB were more affected by the fuel soaking in comparison to BG modified binder.
7. Master curve results also proved that BG modification significantly improved high temperature characteristics of the neat binder such as the rut resistance and the elastic behavior of the binder.

References

Brown, E.R., et al.: Hot Mix Asphalt Materials, Mixtures Design, and Construction, 3rd edn. NAPA Research and Education Foundation, Lanham, MD (2009). ISBN 0-914313-02-1

Cecil, R., Louis, C.F.: Protein-hydrocarbon interactions. Interactions of various proteins with decane in the presence of alcohols. Biochem. J. **117**, 147–156 (1970)

Christensen, D.W., Anderson, D.A.: Interpretation of dynamic mechanical test data for paving grade asphalt. J. Assoc. Asphalt Paving Technol. **61**, 67–116 (1992)

Corun, R., Van Rooijen, R.C., de Bondt, A.H.: Performance Evaluation of Jet Fuel Resistant Polymer-Modified Asphalt for Airport Pavements. American Society of Civil Engineers (2006)

Dehouche, N., Kaci, M., Mokhtar, K.A.: Influence of thermo-oxidative aging on chemical composition and physical properties of polymer modified bitumens. Constr. Build. Mater. **26**, 350–356 (2012)

Li, Y., Lyngdal, E., Bahia, H.: Effect of polymer and oil modification on the aging susceptibility of asphalt binders. Transportation Research Record: Journal of the Transportation Research Board, Washington, DC, pp. 28–37 (2016)

Nick, P.C., Saul, T., Prabhakaran, E.N., Martin Scholtz, J.: Protein structure, stability and solubility in water and other solvents. Philos. Trans. R. Soc. B Biol. Sci. **359**(1448), 1225–1235 (2004)

Rizvi, H.R., Khattak, M.J., Gallo, A.A.: Bone Glue Modified Asphalt: A Step Towards Energy Conservation and Environment Friendly Modified Asphalts. International Scholarly Research Notices, vol. 2014, p. 5. Article ID 807043 (2014)

Rizvi, H.R., Khattak, M.J., Gallo, A.A.: Rheological and mechanistic characteristics of bone glue modified asphalt binders. Constr. Build. Mater. **88**, 64–73 (2015)

Rooijen, R.C., De Bondt, A.H., Corun, R.L.: In: 2004 FAA Worldwide Airport Transfer Conference, Atlantic City, NJ (2004)

Shoenberger, J.S.: Fuel-Resistant Coating and Binders for Porous Friction Surface Pavements: Tests and Analyses. Engineering & Services Laboratory Air Force Engineering & Services Center, Tyndall Air Force Base, FL. ESL-TR-83-33 (1983)

Evaluating the Thermal Characteristics of Rubberized Asphalt by Applying the Object-Based Approach

Shenming Wu[1], Yishuo Huang[1(✉)], Yu-Min Su[2], and Yuan-Zhih Lin[1]

[1] Department of Construction Engineering, Chaoyang University of Technology, Taichung, Taiwan
yishuo@cyut.edu.tw
[2] Department of Civil Engineering, National Kaohsiung University of Applied Sciences, Kaohsiung, Taiwan

Abstract. Rubberized asphalt concrete (RAC) is the pavement material containing asphalt concrete and the crumbed rubber made from the recycled tires. RAC provides an environmentally friendly alternative to pavement materials. Thermal segregation caused by temperature differences presented in hot mix asphalt (HMA) can be used as defect indicators of physical segregation, irregularities, and low densities. The surface temperature information recorded by a thermal camera can usually be used as a clue to identify defects shown on the pavement. Thermography can be used to record the surface temperature information. PAVE-IR mounted on a paver is used to monitor the surface temperature of the pavement in real time. However, a roller is employed to visit the pavement after the paver passed over the pavement. A thermal infrared camera can immediately have applied to record the surface temperature information after the roller passed over the pavement. The study proposes an object-based approach to group those pixels with the similar surface temperatures such that each thermography can be automatically composed of the limited regions and the surface temperature distribution in each region is replaced by the average pixel value of the region. In doing so, the regional boundaries can be extracted, and the thermal segregation illustrated on the constructed pavement can be identified after RAC and HMA installed on a mat. Recently, the RAC pavement was introduced into Taiwan, but its related studies are few. This study did collect two kinds of field data: one is to immediately collect the thermal infrared images after RAC installed in a mat, and another one is to immediately collect a series of thermal infrared image HMA installed in a mat. The proposed approach offers an efficient and robust way to analyze the collected thermography, and the analyzed results reveal the important clues to identify the thermal segregation after RAC and HMA installed in mats, separately. By analyzing the collected thermal infrared images, the cooling tendencies can be quantitatively described.

K. Zhang et al. (eds.), Testing and Characterization of Asphalt Materials and Pavement Structures, Sustainable Civil Infrastructures,
https://doi.org/10.1007/978-3-319-95789-0_2

1 Introduction

Remote sensing is a science to retrieve the information about an object through analyzing the data acquired by a device that is not in physical contact with the object (Lillesand and Kiefer 2015). Infrared thermography is the product of remote sensing techniques and is usually used to record the surface temperature of the observed objects. Infrared thermography is widely used to identify defects presented in objects by observing the radiant heat pattern (Huang and Wu 2010). For a thermal camera, radiation is converted to temperature depending on the emissivity value of the specimen. Emissivity values are in the range of 0 and 1, and they are defined as the ratio of a body's emission spectrum at a given temperature to that of a blackbody at the same temperature (Plotnikov and Winfree 1998). If the emissivity value is 1, it means that the test body is a black body; if it is 0, the test body completely reflects all the received energy. Infrared thermal cameras are designed to record the temperatures transformed from emissivity values. Changes in the recorded temperatures on thermal images reveal possible surface flaws (Maldague 2001). Furthermore, those areas in which the surface temperature information is different with their surrounding neighborhoods can be used to indicate the locations of the potential defects presented on objects.

Thermal segregation is a phenomenon caused by lack of temperature homogeneity in the hot mix asphalt (HMA) constituents of the in-place mat of such a magnitude that there is a reasonable expectation of accelerated pavement distresses (Stroup-Gardiner and Brown 2000); in another word, the surface temperature information illustrated on the mat cannot be distributed evenly. Thermal segregation can be classified into two groups: aggregate segregation and thermal segregation, and both of them can reduce pavement service life (Song et al. 2009). Aggregate segregation usually can be identified and observed by employing human visual inspection because of different textures shown on asphalt mats. As for thermal segregation, it could be caused by placing a cooler mass into the pavement or observing large temperature differentials because of insufficient compaction during pavement installation (Adams et al. 2001; Read 1996). It is difficult to locate those areas with large temperature differentials by employing human visual inspection. Thermography can accurately provide the surface temperature information of a large area. Recently, Texas Transportation Institute (TTI) has developed PAVE-IR to monitor the mat surface temperature of the HMA being placed. PAVE-IR is used to record the mat surface temperature in real-time such that those areas with lower surface temperature can be identified immediately and fixed by placing new HMA from the thermal profile generated by PAVE-IR. Texas Department of Transportation (TxDOT) applies the thermal profile in the construction specifications. However, those locations identified with thermal segregation still remain anomalies in the mat (Sebesta and Scullion 2012).

The paper focuses on analyzing the thermal infrared images taking after a mat had been rolled. The surface temperature information recorded in thermography provides an important clue to identifying the locations of thermal segregation. Image segmentation was employed to segment the collected thermal images such that regional boundaries of the segmented regions can be determined. Traditionally, a penetrating thermometer is employed to measure the surface temperature information on the few sampled points. However, the surface temperature information retrieved by employing a penetrating thermometer can only provide little information of thermal segregation occurring on HMA pavement construction because the limited points were measured by the thermometer. Thermography not only provides the surface temperature information of thermal segregation but also offers enough sampled points to reveal the cooling conditions of pavement construction. Several researchers have proposed different algorithms to analyze thermography by applying image processing techniques: Weritz et al. analyzed the thermal images with Fourier transform to obtain the phase image in the frequency domain (Weritz et al. 2005); Abdel-Qader et al. used segmentation techniques to isolate defects near a concrete bridge deck (Abdel-Qader et al. 2008); Huang et al. applied multilayer segmentation to group those pixels with similar surface temperature information such that the whole thermography can be divided into several regions and the temperature distributions in the segmented regions are homogeneous (Huang et al. 2014). Furthermore, the segmented regions not only quantitatively give the surface temperature information but also identify the regional boundaries such that the locations of uneven surface temperature distributions can be determined. This paper applies the image segmentation on RAC and HMA to identify those possible areas with thermal segregation, and their surface temperature changes during cooling status after pavement constructions.

The remainder of this paper is organized as follows. In the next section, the multilayer level set model is briefly introduced. In Sect. 3 illustrates the segmented results by employing the multilayer level set approach on real thermal infrared images. Eventually, some conclusions and related discussions are provided.

2 Segmented Thermography by Employing the Multilayer Segmentation

The fundamental idea of applying image segmentation on thermography is to partition a given thermal infrared image into a series of sub-regions such that each sub-region is homogeneous. In doing so, the surface temperature distributions can be identified and located in the thermal infrared image. Multilayer segmentation proposed by Chung and Vese segments the given thermal infrared image according to the pre-defined thresholds such that the pixel values shown in the segmented regions are grouped and centered by the pre-selected thresholds (Chung and Vese 2010). The multilayer segmentation is numerically implemented by level set functions proposed by Osher and Fedkiw (2002). In the multilayer approach, let I_0 be a thermal image and the two level set functions are used to segment the image into a series of sub-regions with distinct level values $\{l_1 < l_2 < \cdots < l_m\}$ and $\{k_1 < k_2 < \cdots < k_n\}$ for ϕ_1 and ϕ_2 the level set functions, respectively. Then the energy functions generated by level set functions for ϕ_1 and ϕ_2, pre-selected thresholds $\{l_1 < l_2 < \cdots < l_m\}$ and $\{k_1 < k_2 < \cdots < k_n\}$, and the regional constants c (the average values of the segmented regions) can be defined as follows (Chung and Vese 2010):

$$
\begin{aligned}
E(c,\phi) = & \sum_{,i,j=1}^{m-1,n-1} \int_{\Omega} \left|\mathrm{I}_0 - c_{i,j}\right|^2 \chi_1 dx + \sum_{i=1}^{m-1} \int_{\Omega} \left|\mathrm{I}_0 - c_{\mathrm{i},0}\right|^2 \chi_2 dx \\
& + \sum_{i=1}^{m-1} \int_{\Omega} \left|\mathrm{I}_0 - c_{\mathrm{i,n}}\right|^2 \chi_3 dx + \sum_{j=1}^{n-1} \int_{\Omega} \left|\mathrm{I}_0 - c_{0,\mathrm{j}}\right|^2 \chi_4 dx \\
& + \sum_{j=1}^{n-1} \int_{\Omega} \left|\mathrm{I}_0 - c_{\mathrm{m,j}}\right|^2 \chi_5 dx + \int_{\Omega} \left|\mathrm{I}_0 - c_{0,0}\right|^2 \chi_6 dx + \int_{\Omega} \left|\mathrm{I}_0 - c_{0,\mathrm{n}}\right|^2 \chi_7 dx \\
& + \int_{\Omega} \left|\mathrm{I}_0 - c_{\mathrm{m},0}\right|^2 \chi_8 dx + \int_{\Omega} \left|\mathrm{I}_0 - c_{\mathrm{m,n}}\right|^2 \chi_9 dx \\
& + \mu \sum_{i=1}^{m} \int_{\Omega} \left|\nabla H(\phi_1 - l_i)\right| dx + \mu \sum_{j=1}^{n} \int_{\Omega} \left|\nabla H(\phi_2(x) - k_j)\right| dx
\end{aligned} \tag{1}
$$

where H is the Heaviside function, $\mu > 0$ is a weight parameter, and c_{ij} is the sub-regional constant. The parameter χ_i is the combinations of level set functions ϕ_1 and ϕ_2; for an example, χ_1 is defined as $H(\phi_1 - l_i)H(l_{i+1} - \phi_1)H(\phi_2 - k_j)$ $H(k_{j+1} - \phi_2)$. As for other parameters χ_i, those parameters can be referred the paper of Huang et al. (2014). The optimal approximation I derived from the segmented results can be shown as:

$$
\begin{aligned}
\mathrm{I} = & \sum_{i=1}^{m-1} \sum_{j=1}^{n-1} c_{ij}\chi_1 + \sum_{i=1}^{m-1} c_{i0}\chi_2 + \sum_{i=1}^{m-1} c_{in}\chi_3 + \sum_{j=1}^{n-1} c_{0j}\chi_4 \\
& + \sum_{j=1}^{n-1} c_{mj}\chi_5 + c_{00}\chi_6 + c_{0n}\chi_7 + c_{m0}\chi_8 + c_{mn}\chi_9.
\end{aligned} \tag{2}
$$

The level set functions ϕ_1 and ϕ_2 can be found by employing finite difference such that initial level set functions will change their shapes to meet the energy minimization steps by using iteration scheme (Chung and Vese 2010; Huang and Wu 2010; Huang et al. 2014).

3 Processed Results by Employing Multilayer Segmentation

Thermal infrared camera (Thermo Gear-G120, NEC) was immediately used to record a series of thermal infrared images such that the surface temperature information shown on the pavements right after the pavement was rolled. The thermal infrared camera offers 240 pixels × 320 pixels image sizes with a thermal resolution of 0.1 °C. For a RAC pavement, the location of 45K of Taiwan Provincial Highway No. 1 was selected, and a series of thermal images were recorded at 5-min intervals. The recorded date can be tracked back to May 19th, 2015, and the measured times were from 14:35 to 16:03. As for an AC pavement, a county road was chosen, similarly, and a series of thermal images were recorded at 5-min intervals. The recorded date was June 16th, 2015, and the measured times were from 11:03 to 12:26. Totally, there were 19 thermal infrared images for RAC, and 16 thermal infrared images for AC, respectively. In Fig. 1, two images were selected: one is for RAC, and another one is for AC. Then, the multilayer segmentation used in this paper was employed to analyze the given thermal images.

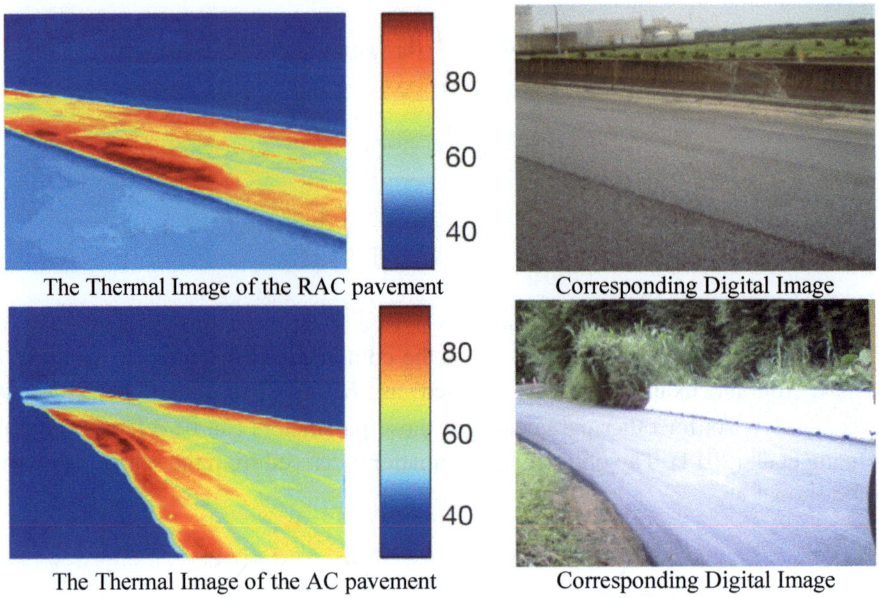

Fig. 1. The thermal images of RAC and HMA pavements (240 by 320 pixels), and their corresponding digital images (960 by 1280 pixels)

The multilayer segmentation is implemented by employing two initial level set functions ϕ_1 and ϕ_2; the level set function ϕ_1 is defined as follows:

$$\phi_1(x, y) = \sqrt{(x - x_i)^2 + (y - y_i)^2} - r \tag{3}$$

where r is the specified radius. As for the level set function ϕ_2, it is generated by Eq. (3) with several pixels of the centers of the function ϕ_1. For simplicity, the pre-selected thresholds were chosen as the same values for the level set functions ϕ_1 and ϕ_2. The surface temperature information is transformed to the digital numbers between 0 and 255. In doing so, the thresholds are set as 0, 20, 50, 70, 100, 150, 200 and 230. The parameter $\mu = 0.005 \times 256 \times 256$ is selected. With employing those parameters into the multilayer segmentation, the segmented results are classified by five classes. The average temperature value of each segmented region is used to replace pixel values contained in the segmented regions. The same thresholds and parameters are applied on the RAC and HMA pavements.

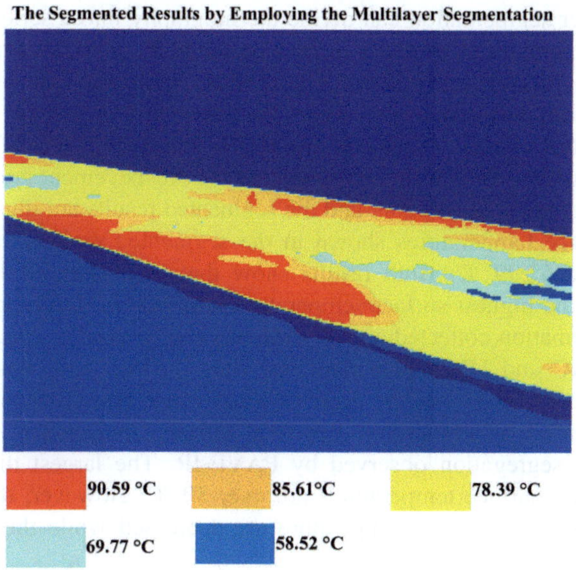

Fig. 2. The segmented results by employing the multilayer segmentation on the RAC pavement, and the thermal image can be grouped into five classes; the average surface temperatures of the segmented regions are 90.59, 85.61, 78.39, 69.77 and 58.52 °C

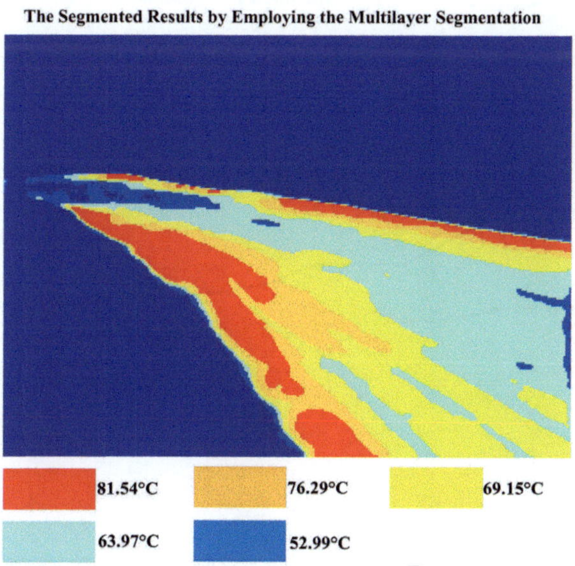

Fig. 3. The segmented results by employing the multilayer segmentation on the HMA pavement, and the thermal image can be grouped into five classes; the average surface temperatures of the segmented regions are 81.54, 76.29, 69.15, 63.97 and 52.99 °C

The multilayer segmentation can divide thermography into a series of sub-regions such that the temperature distribution in each segmented region is homogeneous. With employing the multilayer segmentation on the RAC pavements, the segmented results of the first thermal infrared image were illustrated in Fig. 2; the average surface temperatures of the segmented regions are 90.59, 85.61, 78.39, 69.77 and 58.52 °C, respectively. Similarly, the 7th thermal image of the AC pavement are used to analyze the surface temperature distribution, and the segmented results are illustrated in Fig. 3; the average surface temperatures shown in the segmented regions are 81.54, 76.29, 69.15, 63.97 and 52.99 °C. Both results show that both sides of RAC and HMA pavements have the highest surface temperature information. In doing so, the surface temperature information collected at every interval can be used to describe the cooling behaviors of RAC and HMA pavement materials. In Fig. 4, it was found that the surface temperature of RAC will be decreased faster than HMA. However, the cooling phenomena observed by the analyzed thermal infrared images are more complicate than the thermal segregation observed by PAVE-IR. The largest differences of the segmented average surface temperatures are over 30 °C. However, the differences of the segmented average surface temperature are decreased while the cooling time is increased.

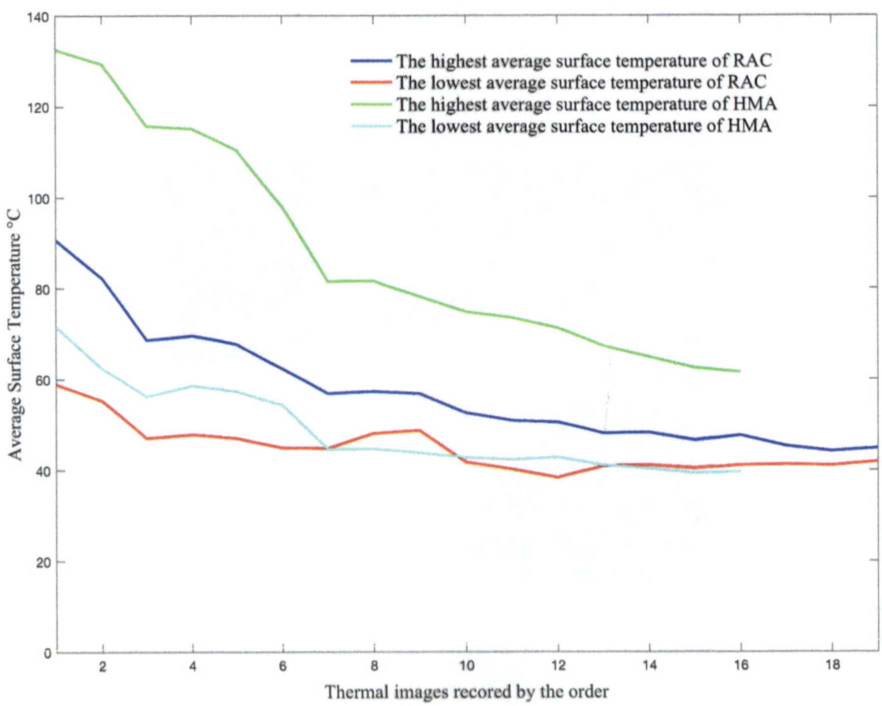

Fig. 4. The tendencies of the highest and lowest average surface temperatures of the RAC and HMA

4 Conclusions

Thermography can be used to illustrating the surface temperature information; the multilayer segmentation can be used to analyze thermography and provide more quantitative descriptions for surface temperature distributions. From the segmented results, the thermal images can be divided into several groups by the surface temperature distributions. In doing so, the surface temperature differentials can be determined and located. PAVE-IR is employed to monitor the thermal segregation during pavement construction, and those areas with thermal segregation will be fixed. However, a thermal infrared camera can be immediately installed to monitor the surface temperature distributions of the rolled pavement. Image segmentation proposed in this paper can automatically group those pixels with the similar surface temperatures such that the surface temperature distributions can be identified. The distributions could be the clue to explain why those repaired areas with thermal segregation still have anomalies.

Acknowledgements. The authors like to show their appreciations for the finical support from Taiwan Construction Research Institute such that the work can be done.

References

Abdel-Qader, I., Yohali, S., Abudayyeh, O., Yehia, S.: Segmentation of thermal image for non-destructive evaluation of bridge decks. NDT&E Int. **41**, pp. 395–405 (2008). https://doi.org/10.1016/j.ndteint.2007.12.003 (Elsevier)

Adams, J., Mulvaney, R., Reprovich, B., Worel, B.: Investigation of Construction-Related Asphalt Concrete Pavement Temperature Differentials. Minnesota Department of Transportation, St. Paul, MN 55155, USA (2001)

Chung, G., Vese, L.A.: Image segmentation using a multilayer level-set approach. Comput. Vis. Sci. **12**(6), 267–285 (2010). https://doi.org/10.1007/s00791-008-0113-1

Huang, Y., Wu, J.: Infrared thermal image segmentations employing the multilayer level set method for non-destructive evaluation of layered structures. NDT&E Int. **43**(1), 34–44 (2010). https://doi.org/10.1016/j.ndteint.2009.08.001

Huang, Y., Chiu, C.T., Lee, M.G., Lin, S.Y.: Applying the multilayer level set approach to explore the thermal surface characteristics of hot mix asphalt. Constr. Build. Mater. **53**, 621–634 (2014). https://doi.org/10.1016/j.conbuildmat.2013.10.097

Lillesand, T.M., Kiefer, R.W.: Remote Sensing and Image Interpretation, 7th edn. Wiley, New York (2015)

Maldague, X.P.V.: Theory and practice of infrared technology for non-destructive testing. Wiley, New York (2001)

Osher, S., Fedkiw, R.: Level Set Methods and Dynamic Implicit Surfaces. Springer, Berlin (2002)

Plotnikov, Y.A., Winfree, W.P.: Advanced image processing for defect visualization in infrared thermography. In: Proceedings of SPIE, The International Society for Optical Engineering, Bellingham, WA, USA, vol. 3361, pp. 331–338 (1998)

Read, S.A.: Construction related temperature differential damage in asphalt concrete pavements. Master's Thesis, University of Washington, Seattle, WA (1996)

Sebesta, S., Scullion, T.: Performance monitoring pavements with thermal segregation in Texas. Report 0-6080-1, Texas Transport Institute, College Station, Texas, USA (2012)

Song, J., Abdelrahman, M., Asa, E.: Use of a thermal camera during asphalt pavement construction. Final report submitted to North Dakota Department of Transportation, Bismarck, ND 58504, USA (2009)

Stroup-Gardiner, M., Brown, R.R.: Segregation in hot mix asphalt pavements. National Cooperative Highway Research Program Report 441. Transportation Research Board, National Research Council, Washington, D.C (2000)

Weritz, F., Arndt, R., Röllig, M., Maierhofer, C., Wiggenhauser, H.: Investigation of concrete structures with pulse phase thermography. Mater. Struct. **38**, 843–849 (2005)

Effect of Anti-stripping Agents on Asphalt Mix Performance Using a Mechanistic Approach

Shivani Rani[1(✉)], Rouzbeh Ghabchi[2], Syed Ashik Ali[1], and Musharraf Zaman[1]

[1] University of Oklahoma Norman, Norman, OK, USA
{shivani.rani, syed.a.ali, zaman}@ou.edu
[2] South Dakota State University Brookings, Brookings, SD, USA
Rouzbeh.Ghabchi@sdstate.edu

Abstract. Anti-Stripping Agents (ASAs) are widely used by the asphalt industry to improve the resistance of asphalt mixes to moisture-induced damage. The effectiveness of an ASA largely depends on its type, asphalt binder type and source, binder content, and aggregate type. This study was undertaken to evaluate the effects of two types of ASA on the rheology, Superpave® Performance Grade (PG) and moisture-induced damage potential of a PG 58-28 asphalt binder. For this purpose, the asphalt binder was blended with 0.5% of two different types of ASA using a high shear mixer. Surface Free Energy (SFE) components of the binder blends were determined and used to evaluate work of adhesion, work of debonding and energy ratio values of the asphalt-aggregate (limestone and granite) systems to assess their moisture-induced damage potential. It was found that incorporation of the ASA in the binder slightly increased the rutting resistance. However, no significant changes in the dynamic viscosity were observed as a result of adding ASA to the binder. The SFE results revealed that adding both types of ASA to the binder improved the bonding between the binder and a granite aggregate, indicating a reduced moisture-induced damage potential, as expected. The moisture-induced damage potential of the PG 58-28 binder with a limestone aggregate was found to reduce by using an amine-based ASA. The findings of this study demonstrate that the SFE-based approach is an effective tool for the assessment of moisture-induced damage potential of asphalt mixes.

Keywords: Anti-Stripping agent · Superpave® performance grading Rutting · Dynamic viscosity · Surface free energy · Moisture-Induced damage And dynamic contact angle

1 Introduction

Moisture-induced damage in asphalt pavements is a serious concern as it leads to localized bleeding, particle disintegration, shoving, potholes or structural failure (Kennedy et al. 1983; Brown et al. 2009; Apeagyei et al. 2014). Moisture-induced damage usually occurs due to lack of either the cohesion resistance of the asphalt binder or the adhesion between aggregate and asphalt binder which allows water to penetrate in the asphalt binder-aggregate interface and reduce its strength (Hicks 1991).

K. Zhang et al. (eds.), *Testing and Characterization of Asphalt Materials and Pavement Structures*, Sustainable Civil Infrastructures,
https://doi.org/10.1007/978-3-319-95789-0_3

Taylor and Khosla (1983) specified five different mechanisms for adhesion bond failure and stripping of asphalt pavements. These mechanisms are: detachment, displacement, spontaneous emulsion (formation of inverted emulsion), increased pore pressure due to reduction in air voids, and hydraulic scouring under compression and tension cycles. According to Harnish (2010), development of negative charge in the presence of water at the surface of asphalt binder (Carboxylic and water interaction) and aggregate (silanol and water interaction) creates repulsive forces at their interface, which leads to stripping. To enhance the adhesion between the asphalt binder and aggregate or to reduce the moisture-induced damage susceptibility, use of an anti-stripping agent (ASA), such as hydrated lime, has been found to be successful (Divito and Morris 1982; Curtis 1990; Solaimanian et al. 1993; Gore 2003; Abo-Qudais and Mulqi 2005; Lu and Harvey 2006; Hossain et al. 2011; Al-Qadi et al. 2014; Souliman et al. 2014). However, the effectiveness of an ASA largely depends on the ASA type and amount used, asphalt binder type and source, binder content, and aggregate type (Pickering et al. 1992; Hossain and Zaman 2009).

Evaluation of the effectiveness of an ASA in reducing moisture-induced damage potential of asphalt mixes in the laboratory is generally based on Tensile Strength Ratio (TSR) test, Lottman Test, Hamburg wheel tracking test, and boiling water test (Divito and Morris 1982; Al-Qadi et al. 2014; Souliman et al. 2014). None of these tests, however, incorporates the damage mechanisms (debonding of asphalt binder and aggregate) noted previously. Therefore, many researchers (see e.g., Cheng et al. 2002; Hefer et al. 2006; Bhasin et al. 2007; Wasiuddin et al. 2007; Ghabchi et al. 2014) have employed a mechanistic approach, namely Surface Free Energy (SFE), to evaluate the moisture-induced damage potential of asphalt mixes. In the SFE technique, moisture-induced damage potential of an asphalt-aggregate system is evaluated by measuring the SFE components of both asphalt binder and aggregate and determining the energy ratio. Birgisson et al. (2004) used the SFE approach and concluded that energy ratio can detect the moisture-induced damage potential of an asphalt mix, in presence of an ASA. The present study was undertaken to evaluate the effects of two types of ASAs on the moisture-induced damage potential of a commonly used PG 58-28 asphalt binder. Also, the Superpave® asphalt binder tests were conducted following applicable standards (AASHTO M 320 2011).

2 Materials and Methodology

2.1 Materials

For this study, an asphalt binder, labeled as PG 58-28, was collected from an Oklahoma refinery. Also, two different types of amine-based ASAs were collected from two different sources in Oklahoma, referred to as ASA1 and ASA2. According to manufacturers, both additives have specific gravities close to 1.0 and are expected to improve the adhesion between the asphalt binder and aggregate.

2.2 Sample Preparation

To evaluate the effects of ASAs on moisture-induced damage of an asphalt mix, the collected binder was blended with 0.5% of ASA1 and ASA2 using a high shear mixer. The rotational speed of the mixer was selected as 1000 rpm and the mixing was performed for 45 min at 145 °C. To evaluate the effect of the short-term and long-term aging on the asphalt binder blends, Rolling Thin Film Oven (RTFO) and Pressure Aging Vessel (PAV) were also performed on asphalt binder samples following AASHTO T 240 (AASHTO 2011) and AASHTO R 28 (AASHTO 2011) standard test methods, respectively.

2.3 Dynamic Shear Rheometer

The rheological properties of the binder blends, namely complex modulus (G^*) and phase angle (δ), were measured at high in-service temperatures (55°, 58°, and 61 °C) and at intermediate in-service temperatures (16°, 19°, and 22 °C) using a Dynamic Shear Rheometer (DSR), in accordance to AASHTO T 315 (AASHTO 2011) standard. These temperatures were selected based on the Performance Grade (PG) of the base binder, namely high PG −3, high PG, high PG +3. The results were analyzed to determine rutting factor ($G^*/\sin\delta$) and Superpave® high-temperature Performance Grade (PG) of the binder blends. To determine the high-temperature PG, a linear regression analysis was conducted on the $G^*/\sin\delta$ values at the selected test temperatures. The lower of two temperatures at which $G^*/\sin\delta = 1.0$ kPa for unaged condition or $G^*/\sin\delta = 2.2$ kPa for RTFO-aged condition was selected as the high-temperature PG of the binder. Similarly, the results at intermediate in-service temperatures were used to calculate the fatigue factor ($G^*.\sin\delta$).

2.4 Bending Beam Rheometer

The low-temperature grading parameters, namely stiffness (S) and creep relaxation (m), of the binder blends were measured by conducting Bending Beam Rheometer (BBR) test in accordance to AASHTO T 313 (AASHTO 2011). The tests were conducted at −18° and −21 °C on PAV-aged asphalt binders. Similar to high-temperatures, the low test temperatures were selected as low PG and low PG −3.

2.5 Rotational Viscometer

Rotational Viscometer (RV) tests were conducted in accordance to AASHTO T 316 (AASHTO 2011) standard test method to determine the dynamic viscosity of the binder blends. For this study, the tests were conducted at four different temperatures, namely 135°, 150°, 165°, and 180 °C and at 20 rpm shearing rate. Three consecutive readings were taken manually at one-minute interval once the temperature reached the set temperature and viscosity value on the display became constant.

2.6 Surface Free Energy

The SFE components, namely acid SFE (Γ^{+}), base SFE (Γ^{-}), acid-base SFE $\left(\Gamma^{AB}\right)$, Lifshitz-van der Waals (Γ^{LW}), and total SFE components of the binder blends were determined from the Dynamic Contact Angles (DCAs) measured using the dynamic Wilhelmy plate Method. Equations (1)–(3) were used to determine the SFE components based on the measured DCA values of the binder with selected probe liquids. Since Eq. (1) has three unknowns, therefore three different probe liquids, namely water, glycerin, and formamide, were used for testing. As suggested by Bhasin et al. (2007), the work of adhesion (W_{AS}), the work of debonding (W_{ASW}^{wet}), and the energy ratio (ER) were determined using Eqs. (4), (5) and (6), respectively, to calculate the moisture-induced damage potential of asphalt-aggregate systems.

$$\Gamma_L(1+\cos\theta) = 2\left(\sqrt{\Gamma_A^{LW}\Gamma_L^{LW}} + \sqrt{\Gamma_A^{+}\Gamma_L^{-}} + \sqrt{\Gamma_A^{-}\Gamma_L^{+}}\right) \tag{1}$$

$$\Gamma^{Total} = \Gamma^{LW} + \Gamma^{AB} \tag{2}$$

$$\Gamma^{AB} = 2\sqrt{(\Gamma^{+}\Gamma^{-})} \tag{3}$$

$$W_{AS} = 2\left(\sqrt{\Gamma_A^{LW}\Gamma_S^{LW}} + \sqrt{\Gamma_A^{+}\Gamma_S^{-}} + \sqrt{\Gamma_A^{-}\Gamma_S^{+}}\right) \tag{4}$$

$$W_{ASW}^{wet} = \Gamma_{AW} + \Gamma_{SW} - \Gamma_{AS} \tag{5}$$

$$ER = \left|\frac{W_{AS}}{W_{ASW}^{wet}}\right| \tag{6}$$

where, subscripts A, S, W, and L represent the parameters corresponding to the binder, stone (or aggregate), water, and probe liquid, respectively. Therefore, parameters Γ_{AW}, Γ_{SW}, and Γ_{AS} defines the interfacial energy between binder and water, aggregate and water, and binder and aggregate, respectively. θ is the dynamic contact angle between the binder and the probe liquid.

3 Results and Discussion

3.1 Performance at High Temperatures

The rutting factors of the binder blends measured at high in-service temperatures are summarized in Fig. 1. The rutting factor was measured for three different samples and the standard deviations of the results are also presented in Fig. 1. From Fig. 1, the addition of ASAs to binder blends slightly increased the rutting factor of the binder blends. This increase was more noticeable (40%) when ASA1 was used as compared to ASA2 (17%). It is well known that the higher the rutting factor the higher the rutting resistance. Therefore, one can say that the binder containing ASA1 is expected to have the highest rutting resistance followed by the binder containing ASA1 and the neat binder. After RTFO-aging, both neat binder and the blend containing 0.5% ASA2 were found to have similar rutting factors, while the binder contacting ASA1 had 20% higher rutting factor. These results indicated that the use of ASA slightly enhanced the rutting resistance of the binder blend. However, the selection of the ASA type is important as the amount of enhancement depends on the ASA type. Gore (2003) and Xiao and Amirkhanian (2010) also observed an increase in the rutting resistance due to adding ASA in the binder depending on the binder type, source, and type and amount of ASA.

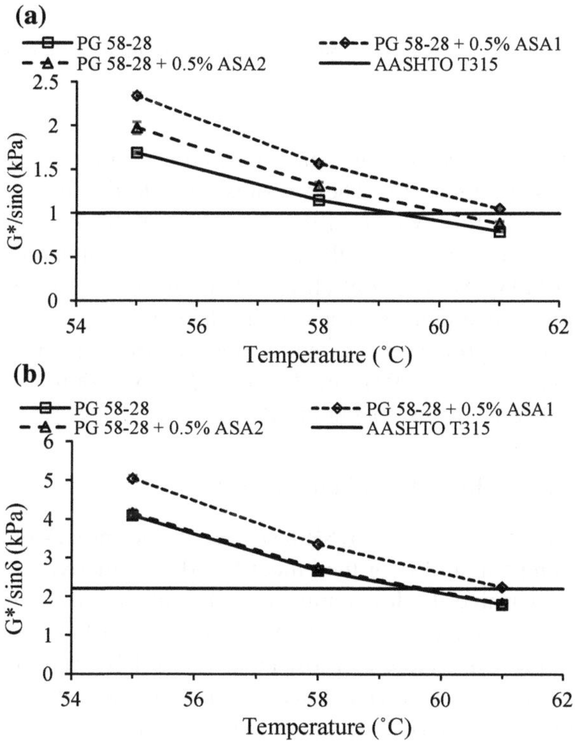

Fig. 1. Rutting factors of unaged (**a**) and RTFO-aged (**b**) asphalt binders measured at high in-service temperatures

3.2 Performance at Intermediate Temperatures

All the asphalt binder blends were tested in a DSR to determine their fatigue factor at intermediate in-service temperatures. The fatigue factor was measured for three different samples and the measured G*.sin δ values and the standard deviations are presented in Fig. 2. As expected, the G*.sin δ values were found to decrease with an increase in temperature. It is well known that the lower the fatigue factor the more durable the binder and the higher the resistance to fatigue cracking. As shown in Fig. 2,

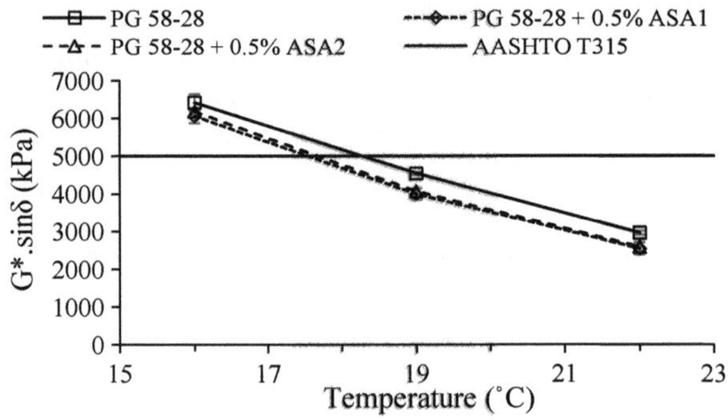

Fig. 2. Fatigue factors of PAV-aged asphalt binders measured at intermediate in-service temperatures

the addition of ASAs, irrespective of their types, reduced the fatigue factor by 10%, indicating an increased fatigue life. These results indicate that the asphalt mixes containing ASA are expected to exhibit an improved long-term performance in the field compared to conventional mixes, without any additive. Similar observations were reported by Gore (2003). Souliman et al. (2014) performed DSR tests on PAV-aged asphalt binder blends containing ASA and found that ASA preserved the elasticity of the binder during long-term oxidative aging and subsequently reduced the fatigue cracking. Solaimanian et al. (1993) and Lu and Harvey (2006) also reported good long-term performance of mixes containing ASA.

3.3 Performance Grading of Asphalt Binders

The measured G*/sin δ values using DSR test were further processed to determine the Superpave® high-temperature PG of the binder blends. Similarly, the BBR test results were analyzed to determine the low-temperature PG of the binder blends. The high- and low-temperature PG grades are presented graphically in Fig. 3. From Fig. 3, it is evident that both high-temperature and low-temperature PG grades of the base binder slightly increased after it was blended with both types of ASAs. Subsequently, a slightly improved rutting performance and reduced low-temperature cracking potential are observed as a result of the addition of ASA. However, none of the ASAs changed the Superpave® PG of the base binder (PG 58-28). Gore (2003), Selvamohan (2007), Xiao and Amirkhanian (2010) and Souliman et al. (2014) also observed no change in the Superpave® PG of the asphalt binders due to the addition of ASA to the binder.

3.4 Dynamic Viscosity of Asphalt Binders

The measured dynamic viscosities of the binder blends at the selected test temperatures are shown in Fig. 4. The dynamic viscosity was measured for three different samples and the standard deviations were calculated, as shown in Fig. 4. From Fig. 4, the addition of ASAs, irrespective of their types, did not significantly affect the dynamic

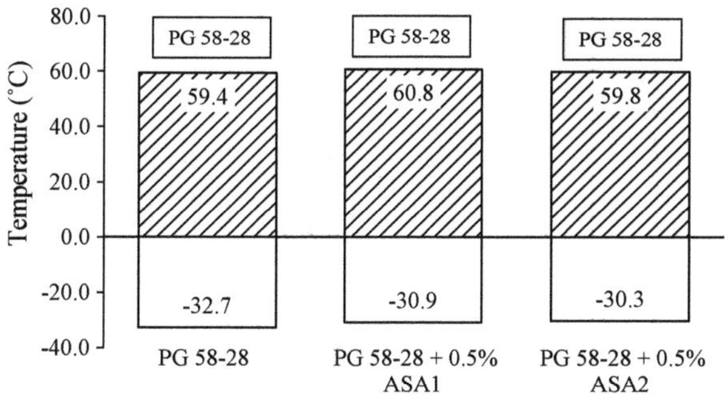

Fig. 3. Continuous and Superpave® performance grade of asphalt binders

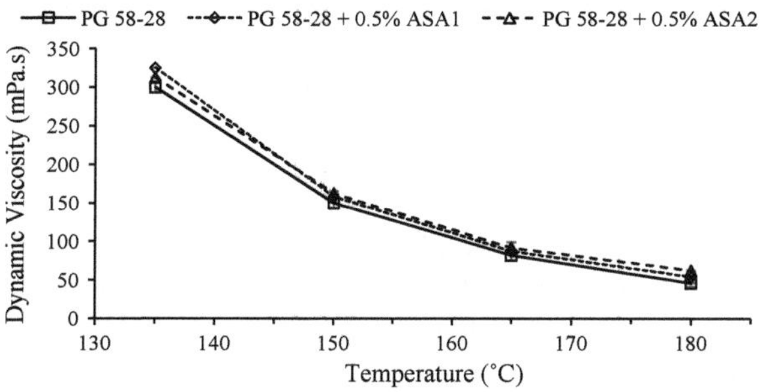

Fig. 4. Dynamic viscosity of asphalt binders at different test temperatures

viscosity of the binder blends. A marginal variation was observed at the lowest test temperature of 135 °C. Gore (2003), Selvamohan (2007), Xiao and Amirkhanian (2010), and Souliman et al. (2014) also reported similar observations.

3.5 Surface Free Energy Components of Asphalt Binders

The calculated DCAs of the binder blends with water, glycerin, and formamide are presented in Table 1 along with their standard deviations. From Table 1, it is evident that the addition of ASA2 did not significantly affect the DCAs of the base binder. However, ASA1 reduced the measured DCA values of the binder blends. A probe liquid can wet the surface of the asphalt binder only if the measured DCA is less than 90° (Shafrin and Zisman 1960). During this study, only the binder containing ASA1 with formamide was found to have a DCA value less than 90.

The calculated SFE components of the binder blends using Eqs. (1), (2), and (3) are presented in Table 1. The results indicated that, in general, the addition of ASA increased the total SFE of the base binder and therefore, it requires a higher energy to break the bond between the binder molecules (Moffat 2003). However, ASA1 was found to increase both acid SFE and base SFE components of the binder. ASA2 did not significantly change the base SFE component, but decreased the acid SFE component of the binder. According to Arabani et al. (2012), a change in the SFE component of the

Table 1. Dynamic contact angle and surface free energy components of the asphalt binders

Binder type	Additive type	Dynamic contact angle (°)					
		Water		Glycerin		Formamide	
		Average	Standard deviation	Average	Standard deviation	Average	Standard deviation
PG 58-28 OK	–	110.2	0.1	99.9	0.1	93.6	0.3
	ASA1	102.7	0.4	93.1	0.6	87.9	0.8
	ASA2	109.4	0.6	99.6	0.3	92.5	0.8
		Surface free energy components (mJ/m^2)					
		Γ^+ (Acid)	Γ^- (Base)	Γ^{LW} (Non-polar)	Γ^{AB}	Γ^{total}	
PG 58-28 OK	–	0.15	0.74	14.09	0.67	14.76	
	ASA1	0.63	2.05	13.47	2.27	15.75	
	ASA2	0.02	0.81	16.64	0.26	16.90	
Limestone	–	17.50	741.40	51.40	227.80	279.20	
Granite		24.10	96.00	133.20	96.20	229.40	

asphalt binder can impact its interaction with an aggregate; however, cannot quantify the moisture-induced damage potential of an asphalt mix. Therefore, the work of adhesion, the work of debonding, and the energy ratio were determined which provides a better estimate of the moisture-induced damage potential than the SFE components alone (Bhasin et al. 2017).

3.6 Work of Adhesion, Work of Debonding, and Energy Ratio

The work of adhesion, work of debonding, and energy ratio values were determined for binder blends with both limestone and granite aggregates, as presented in Fig. 5. The SFE components of the limestone and granite aggregates in Table 1 were adopted from the available literature (Ghabchi et al. 2014; Buddhala et al. 2011). The results indicated that the binders had a higher interface bond strength with granite aggregate because of higher work of adhesion and energy ratio and a lower work of debonding with granite aggregate than those with limestone aggregate. For example, the energy ratio of the neat binder with granite aggregate was found to be 1.8, which is 300% higher than that for limestone aggregate. Therefore, granite aggregates in an asphalt mix may result in a less moisture-induced damage potential compared to mixes containing limestone aggregates. Other researchers have also reported the similar finding (Ghabchi et al. 2014; Moraes et al. 2017). Also, the asphalt binder containing ASA1 had higher work of adhesion (114.3 mJ/m^2) and energy ratio (2.2) and lower work of debonding (53.1 mJ/m^2) with granite aggregate in comparison to other binders. Therefore, the asphalt binder containing ASA1 is expected to have higher adhesion strength at the aggregate and asphalt binder interface and lower moisture-induced damage potential. In summary, the use of ASA1 resulted an improved resistance of asphalt binders to moisture-induced damage with both limestone and granite aggregates. On the other hand, use of ASA2 in binder blends resulted in an increase in resistance of asphalt binder to moisture-induced damage with granite aggregate.

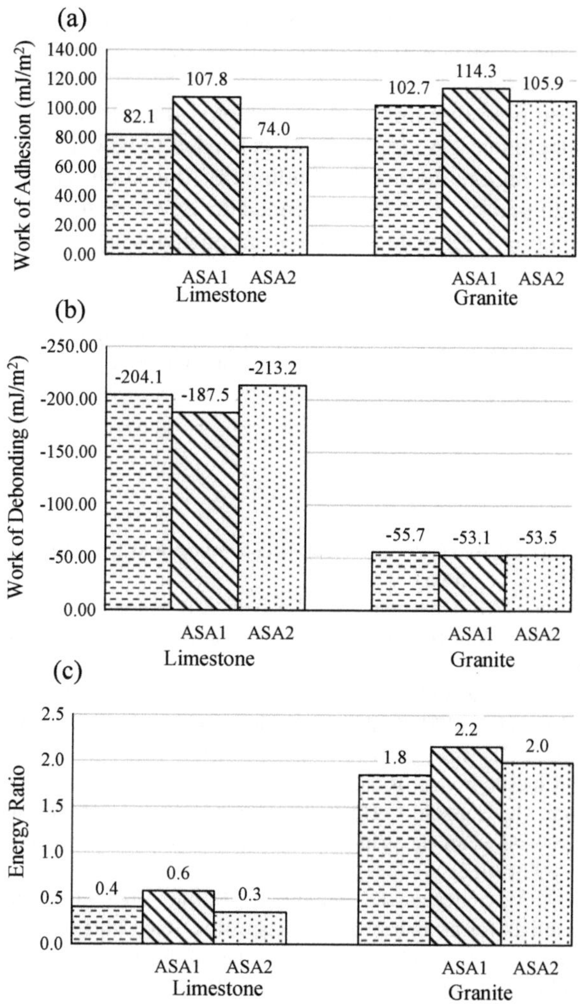

Fig. 5. The work of adhesion (**a**), the work of debonding (**b**), and the energy ratio (**c**) of asphalt binder-aggregate systems

4 Conclusions

This study evaluated the performance of two-different types of ASAs on the rheology, dynamic viscosity, and moisture-induced damage potential of a PG 58-28 asphalt binder. It was found that the addition of ASA to binder slightly increased the high-temperature continuous PG and resistance to rutting of the binder blends. Also, the use of ASA did not affect the dynamic viscosity of the binder. An improvement in the resistance of the selected binders to moisture-induced damage was observed with both limestone and granite aggregates, as a result of using ASA1 in binder blends. However, ASA2 increased the moisture-induced resistance of the binder with granite aggregate. The use of ASA1 was found to be more effective in enhancing the adhesion and increasing of the resistance to moisture-induced damage potential when limestone aggregate is used in a mix.

It is important to note that this study considered one binder, two ASAs, and two different types of aggregates. A similar study is needed to be conducted using the local materials from other regions before the findings of this study can be generalized for other regions. Additionally, it is recommended to develop the master curves of the binders after conducting DSR tests at different frequencies and temperatures in another study.

Acknowledgements. The authors would like to thank the Oklahoma Department of Transportation (ODOT) and the Southern Plain Transportation Center (SPTC) for supporting this research.

References

Abo-Qudais, S., Mulqi, M.W.: New chemical antistripping additives for bituminous mixtures. J. ASTM Int. **2**(8), 1–11 (2005)

Al-Qadi, I.L., Abauwad, I.M., Dhasmana, H., Coenen, A.R.: Effects of Various Asphalt Binder Additives/Modifiers on Moisture-Susceptible Asphaltic Mixtures. Illinois Center for Transportation (2014)

American Association of State Highway and Transportation Officials: Standard Specifications for Transportation Materials and Methods of Sampling and Testing. AASHTO Standards (2011)

Apeagyei, A.K., Grenfell, J.R., Airey, G.D.: Moisture-induced strength degradation of aggregate–asphalt mastic bonds. Road Mater. Pavement Des. **15**(Supp 1), 239–262 (2014)

Arabani, M., Roshani, H., Hamedi, G.H.: Estimating moisture sensitivity of warm mix asphalt modified with Zycosoil as an antistrip agent using surface free energy method. J. Mater. Civ. Eng. **24**(7), 889–897 (2012)

Bhasin, A., Little, D., Vasconcelos, K., Masad, E.: Surface free energy to identify moisture sensitivity of materials for asphalt mixes. Transp. Res. Rec. **2001**, 37–45 (2007)

Birgisson, B., Roque, R., Page, G.: Performance-based fracture criterion for evaluation of moisture susceptibility in hot-mix asphalt. Transp. Res. Rec. **1891**, 55–61 (2004)

Brown, E.R., Kandhal, P.S., Roberts, F.L., Kim, Y.R., Lee, D.Y., Kennedy, T.W.: Hot mix asphalt materials, mixture design and construction, 3rd edn. NAPA Research and Education Foundation, Lanham, Maryland (2009)

Buddhala, A., Hossain, Z., Wasiuddin, N.M., Zaman, M.: Effects of amine anti-stripping agent on moisture susceptibility of Sasobit and Aspha-min mixes by surface free energy analysis. J. Test. Eval. **40**(1), 1–9 (2011)

Cheng, D., Little, D.N., Lytton, R.L., Holste, J.C.: Use of surface free energy properties of the asphalt-aggregate system to predict moisture damage potential. J. Assoc. Asphalt Paving Technol. **71**, 59–88 (2002)

Curtis, C.W.: A literature review of liquid antistripping and test for measuring stripping SHRP—A/UIR–90-016. Strategic Highway Research Program (SHRP). National Research Council, Washington DC (1990)

DiVito, J.A., Morris, G.R.: Silane pretreatment of mineral aggregate to prevent stripping in flexible pavements, Arizona Transportation Research Center, Arizona State University, Phoenix, AZ, USA. Transp. Res. Rec. **843**, 104–111 (1982)

Ghabchi, R., Singh, D., Zaman, M.: Evaluation of moisture susceptibility of asphalt mixes containing RAP and different types of aggregates and asphalt binders using the surface free energy method. Constr. Build. Mater. **73**, 479–489 (2014)

Gore, R.: Effects of amine-based anti-strip additives on the performance grade of asphalt binders. In: Superpave Technical Issue (2003)

Harnish, C.I. (2010). Liquid anti-strip technology & best practices. In: Powerpoint presentation presented at NCAUPG, Overland Park, KS

Hefer, A., Bhasin, A., Little, D.: Bitumen surface energy characterization using a contact angle approach. J. Mater. Civ. Eng. ASCE **18**(6), 759–767 (2006)

Hicks, R.G.: Moisture damage of asphalt concrete. In: National Cooperative Highway Research Program Synthesis, vol. 175 (1991)

Hossain, Z., Zaman, M.: Rheological properties of performance grade binders using a dynamic mechanical analyzer. Pavement. Mater. Model. Test. Perform. 140–149 (2009)

Hossain, Z., Zaman, M., Wasiuddin, N.M., Sneed, J., O'Rear, E.A.: Rheological evaluation of warm mix and anti-stripping additives modified performance grade binders. Road Mater. Pavement Des. **12**(4), 875–895 (2011)

Kennedy, T.W., Roberts, F.L., Lee, K.W.: Evaluation of moisture effects on asphalt concrete mixtures. Transp. Res. Rec. **911**, 134–143 (1983)

Lu, Q., Harvey, J.: Long-term effectiveness of antistripping additives: Laboratory evaluation. Transp. Res. Rec. **1970**, 14–24 (2006)

Moraes, R., Velasquez, R., Bahia, H.: Using bond strength and surface energy to estimate moisture resistance of asphalt-aggregate systems. Constr. Build. Mater. **130**, 156–170 (2017)

Pickering, K., Sebaaly, P.E., Stroup-Gardiner, M., Epps, J.A.: Evaluation of new generation of antistripping additives. Transp. Res. Rec. **1342**, 26 (1992)

Selvamohan, S.: Effect of anti-stripping additives on performance graded binders in Oklahoma. Doctoral dissertation, University of Oklahoma (2007)

Shafrin, E.G., Zisman, W.A.: Constitutive relations in the wetting of low energy surfaces and the theory of the retraction method of preparing monolayers. J. Phys. Chem. **64**(5), 519–524 (1960)

Solaimanian, M., Kennedy, T.W., Elmore, W.E.: Long-term evaluation of stripping and moisture damage in asphalt pavements treated with lime and antistripping agents. Center for Transportation Research, Bureau of Engineering Research, University of Texas at Austin (1993)

Souliman, M.I., Hajj, E.Y., Sebaaly, P.E.: Impact of antistrip additives on the long-term aging rheological properties of asphalt binders. J. Mater. Civ. Eng. **27**(8), C4014006 (2014)

Taylor, M.A., Khosla, N.P.: Stripping of asphalt pavements: state of the art. Transp. Res. Rec. **2001**(911):150–158 (1983)

Wasiuddin, N.M., Fogle, C.M., Zaman, M.M., O'Rear, E.A.: Effect of antistrip additives on surface free energy characteristics of asphalt binders for moisture-induced damage potential. J. Test. Eval. **35**(1), 1–9 (2007)

Xiao, F., Amirkhanian, S.N.: Effects of liquid antistrip additives on rheology and moisture susceptibility of water bearing warm mixtures. Constr. Build. Mater. **24**(9), 1649–1655 (2010)

Stiffening Thin Orthotropic Deck Structures with Thermoset Epoxy Asphalt for Improved Fatigue Resistance

Can Chen[1(✉)], Wolfgang O. Eisenhut[1], Kreisler Lau[1], Alex Ingram[1], and John Bors[2]

[1] ChemCo Systems Inc., Redwood City, USA
{chen, eisenhut, lau}@chemcosystems.com, Alex.Ingram@ahb.nzta.govt.nz
[2] Auckland Harbour Bridge Alliance, Auckland, New Zealand
bors@chemcosystems.com

Abstract. Early fatigue cracking of the pavement was frequently observed on bridges with relatively thin decks of orthotropic design, particularly in hot climates. This study investigates the possible use of epoxy asphalt to stiffen a thin steel deck and mitigate flexural and fatigue failure. Three different epoxy asphalt binders were used and two pavement mix designs were employed. The binder and pavement mixture properties were evaluated with direct tensile test and pull-off strength test on the binder, Marshall and flexural fatigue tests on paving mixture with steel plates. Test results show that a relatively high modulus binder and a relatively stiff pavement can effectively reduce the composite's deflection and improve fatigue resistance. Binder adhesion to the aggregate in the composite seems to play a pivotal role and the strength of this bond may have a close relationship with the stiffness of the pavement.

1 Introduction

For more than 60 years, hundreds of orthotropic steel deck bridges have been successfully employed throughout the world. However, it is recognized that orthotropic bridges have not been problem-free historically. Pavement fatigue cracking has been observed frequently in such decks resulting from the complicated welded details combined with stresses that can be more difficult to quantify and, in particular, early designs which attempted to overly minimize plate thickness to reduce weight (FHWA 2012). Mainly because the inappropriate deflection-to-span requirement for early orthotropic bridge design, approximately 1/500, many bridge deck plates were designed only about 10–12 mm thick, while the stiffening rib spacing are 300 mm (AISC 1963). These may include the Auckland harbor bridge in New Zealand, Port Mann Bridge in Canada, Severin Bridge in England and many else. With rapid increase of vehicle weight and traffic volume nowadays, deflection of these thin deck bridges become higher and flexural fatigue cracking can easily occur on bridge pavement and even influence steel deck and leads to the fatigue of metal (FHWA 2012; Murakoshi et al. 2012). The purpose of this study is to find out the best bridge pavement material

K. Zhang et al. (eds.), *Testing and Characterization of Asphalt Materials and Pavement Structures*, Sustainable Civil Infrastructures,
https://doi.org/10.1007/978-3-319-95789-0_4

that can be used for the rehabilitation of thin deck orthotropic bridges and mitigate fatigue crack failure.

Bridge pavement design for orthotropic steel deck has followed divergent approaches. Guss asphalt system generally are multi-layer structures often in excess of 76 mm thickness using conventional or polymer-modified asphalt binder. The multi-layer structure can hardly meet the weight and density requirement for such older orthotropic bridges, take in account the dead and live load capacity of thin steel deck. America's epoxy asphalt material, on the other approach, is used in this study and the bridge pavement thickness can be designed to about 40 mm or less. Most epoxy asphalt pavement installations have exhibited excellent performance, after years of field performance survey and monitoring (Gaul 1996; Lu and Luo 2010; Jia et al. 2016).

Epoxy asphalt is thermoset, which means that it does not melt at high temperatures. Epoxy asphalt comprises two parts. Part A is epoxy resin and Part B contains asphalt, curing agent and suitable additives. After Part A and Part B are mixed according to a stoichiometric ratio to initiate the chemical reaction which ultimately leads to a cross-linked polymer structure, which can provide high stiffness, strength and adhesion in the composite. The encapsulated asphalt in the polymer structure act as an extender and possibly contributes to the flexibility of the viscoelastic structure. The material not only being used for bridge pavement, modified epoxy asphalt has also been used for pervious pavement, chip seal and other applications (OECD 2017; Alabaster et al. 2008, 2012; Wu et al. 2017).

2 Test Methodology

To the pavement technologist, orthotropic steel decks with their inherent flexibility present a particular challenge. The design of the wearing surface must take into account the flexibility of the deck, pavement mixture stability, density and weight, fatigue under heavy traffic load, as well as the skid characteristics. The following laboratory test methods were used to evaluate the epoxy asphalt and mixture properties:

- Binder direct tensile test, which gives the mechanical characterization of binder in tension.
- Marshall stability test, which is can assess the stability of the pavement immediately after placement and upon full cure.
- Flexural beam fatigue test, measures the effects of repetitive loading on the steel plate and the pavement.
- Pull-off test, measures the binder bond strength to the steel plate and gives an indirect indication to the strength of the bond to the aggregate.

3 Experimentation and Analysis

Asphalt binders do not have a tensile strength in the conventional sense because they are viscous and are in semi-liquid state even in room temperature. Epoxy asphalts become thermoset when the binder is fully cured and the polymer structure is built up. The typical laboratory process to reach full cured condition for epoxy asphalt is at 121 °C for 4 h. At room temperature, it usually takes up to 60–90 days to reach a similar physical condition in the field. Direct tensile strength test—ASTM D 412 usually used in elastomer and rubber were adopted to use in here. Figure 1 shows the full cured epoxy asphalt binder under direct tension at the testing speed of 50 mm/min. Three types of epoxy asphalt binders were used and 6–9 samples were tested for each binder type. Table 1 shows the tensile strength properties of the three epoxy asphalt binders in full cured condition.

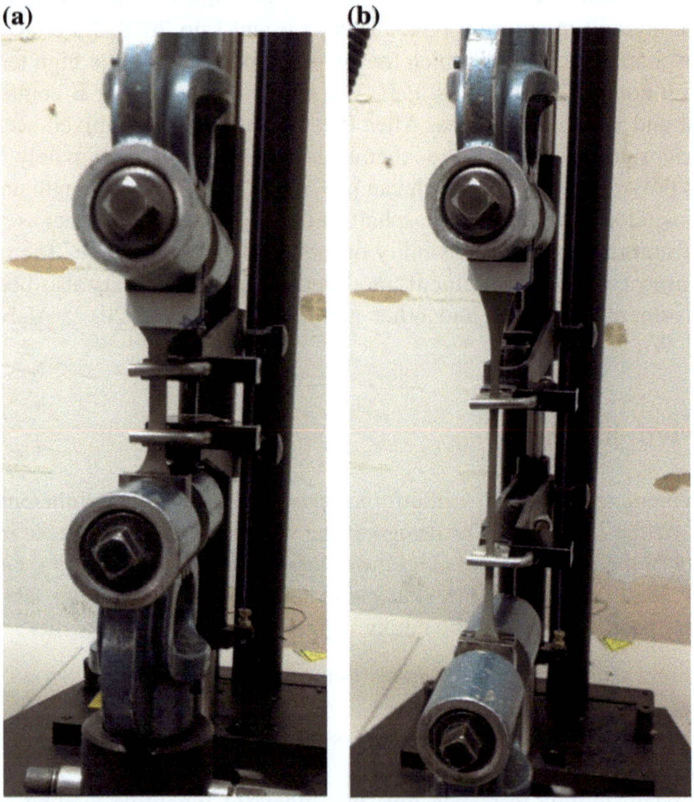

Fig. 1. Rubber and elastomer tensile strength test-ASTM D412 for full-cured epoxy asphalt binder **a** before the tensile test; **b** during the tensile test before rupture

As can be seen, Type A epoxy asphalt is a softer binder with lowest modulus, while Type C binder is much stiffer and higher in molecular weight. However, all three types of epoxy asphalts give similar elongation before rupture under direct tension, which means that although Type B and C binders are much stiffer, they do not lose much flexibility or ductility. Modulus above 100% elongation is not considered, because it is more likely in nonlinear stress-strain relationship and greater strain above 100% probably cannot be accommodated in the mixture under real load condition. All testing

Table 1. Tensile properties of epoxy asphalt in full cured condition

	Type A	Type B	Type C
Tensile strength (MPa)	7.1	10.3	19.4
% Elongation	228.5	248.5	238.5
Modulus @100% elongation (MPa)	1.42	2.98	4.33
Viscosity increase to 1000 cps @ 121 °C (mins)	32	41	54

was conducted at room temperature—23 °C. Time to 1000 cps is the viscosity condition of the epoxy asphalt after mixing and is an indicator of a recommended time length that the epoxy asphalt mixture should be delivered and placed on the bridge deck after production. A shorter time to 1000 cps viscosity usually means a shorter operational time for epoxy asphalt, and above a certain operational window, material could become too cured and harder to work with.

Mix design were attempted for both dense graded (DSG) mix and stone mastic asphalt (SMA) mix with 9.5 mm (3/8-in.) maximum size aggregate, that is, 100% passes the 12.5 mm (1/2 in.) sieve. Aggregate size distribution for the two designs are shown in Fig. 2. It is noted that the DSG gradation is very close and parallel to the maximum density line, while the SMA aggregates have a larger percentage retained on 4.75 mm (#4) sieve. The DSG uses 100% high quality basalt aggregates from China, while the SMA mixture contains 23% lightweight slate crushed fines. The philosophy of SMA is to use high quality coarse aggregates with stone-to-stone contact. However, lightweight aggregates were introduced in this mix to help reduce the density and weight of bridge pavement, and coarse stone mastic structure can provide a rougher surface texture and better skid resistance at the same time. The Marshall Method of Mix Design is used for stability, flow and void analysis for the two types of mixtures, and the optimum binder content were determined based on the lowest air voids of com-

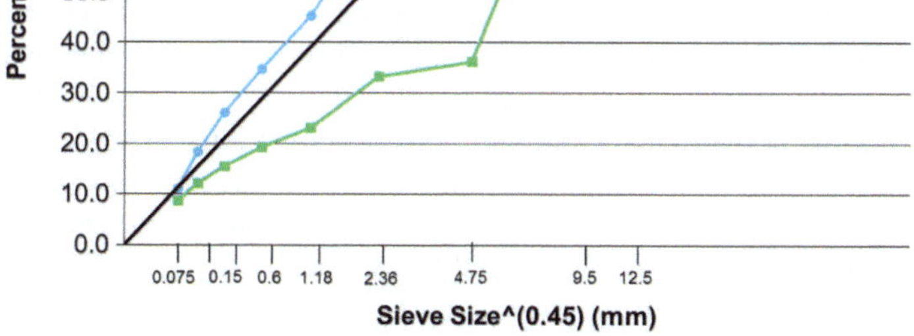

Fig. 2. Aggregate size distribution for dense grade and stone mastic mixture

pacted samples. The optimum binder content for DSG mixture is 6.1% by weight of mix, and SMA is designed rich in binder with 7.5% binder content.

Marshall stability for uncured epoxy asphalt mixture is generally used to estimate when the bridge pavement can be open for traffic and a minimum 8.0 KN is required for uncured Marshall stability according to the MS-2 handbook (Asphalt Institute 1979). As can be seen in Table 2, both mixtures with the three types of binders all meet the stability requirement and the bridge pavement can open to traffic in 12 h after paving. Full cured Marshall stability (with an additional 4 h conditioning in the oven at 121 °C) were not tested, because full cured mixture with Type C binder may easily exceed the stability ring measurement capacity at 90 KN. A minimum of 3 Marshall samples were made, and the density of the DSG samples are around 2.60 g/cm^3 and it is about 2.27 g/cm^3 for SMA with lightweight aggregates added. Both mixtures give air void content less than 3%, in which water permeability and moisture damage do not need to be considered. The density of SMA is about 13% lighter than the DSG and can reduce the deadweight by 13 kg/m^2 based on 40 mm lift thickness of the wearing surface.

Table 2. Marshall values of uncured epoxy asphalt mixtures

		Type A binder	Type B binder	Type C binder
Dense graded mixture	Density (g/cm^3)	2.608	2.593	2.590
	Stability (KN)	11.7	13.1	13.9
Stone mastic mixture	Density (g/cm^3)	2.285	2.275	2.280
	Stability (KN)	8.4	9.8	12.1

Experimentation with the interaction of composite wearing surface and steel deck structure is the most important test content. According to the finite element analysis performed by Seim and Ingham (2004), the curvature of the bridge deck is greater transversely than longitudinally under loading, and stresses are also higher in this direction. Both peak compressive stress and tensile stress would occur on the top fiber of the wearing surface, directly under the load. The peak shear stress is also directly under the load, on the bottom fiber where the wearing surface is bonded to the deck plate. The stress distributions are illustrated in Fig. 3a, b, adopted from Seim and Ingham (2004). To best simulate the stiffener-to-stiffener deflection of real bridge deck plate and wearing surface under load, a simplified small-scale experiment model was designed. A steel plate in 380 mm long by 100 mm wide size and 11 mm thick was fabricated, two stiffeners were welded on each end of the plate and extend out about 40 mm, so that the composite beam can be placed on the test machine firmly. Steel plate span/spacing between two stiffeners are 300 mm long. The schematic of the testing steel plate is shown in Fig. 3c. Composite beam samples were made in three steps: bond coat material was spread on steel plate first, then, hot epoxy asphalt mix were placed onto the bond coat, compacted in a mold and then fully-cured in the oven at 121 °C for 5 h. Finally, extrude the composite beam specimen out of the mold, and ready for the beam fatigue test. Thickness of the mixtures are all 40 mm (±1 mm), and

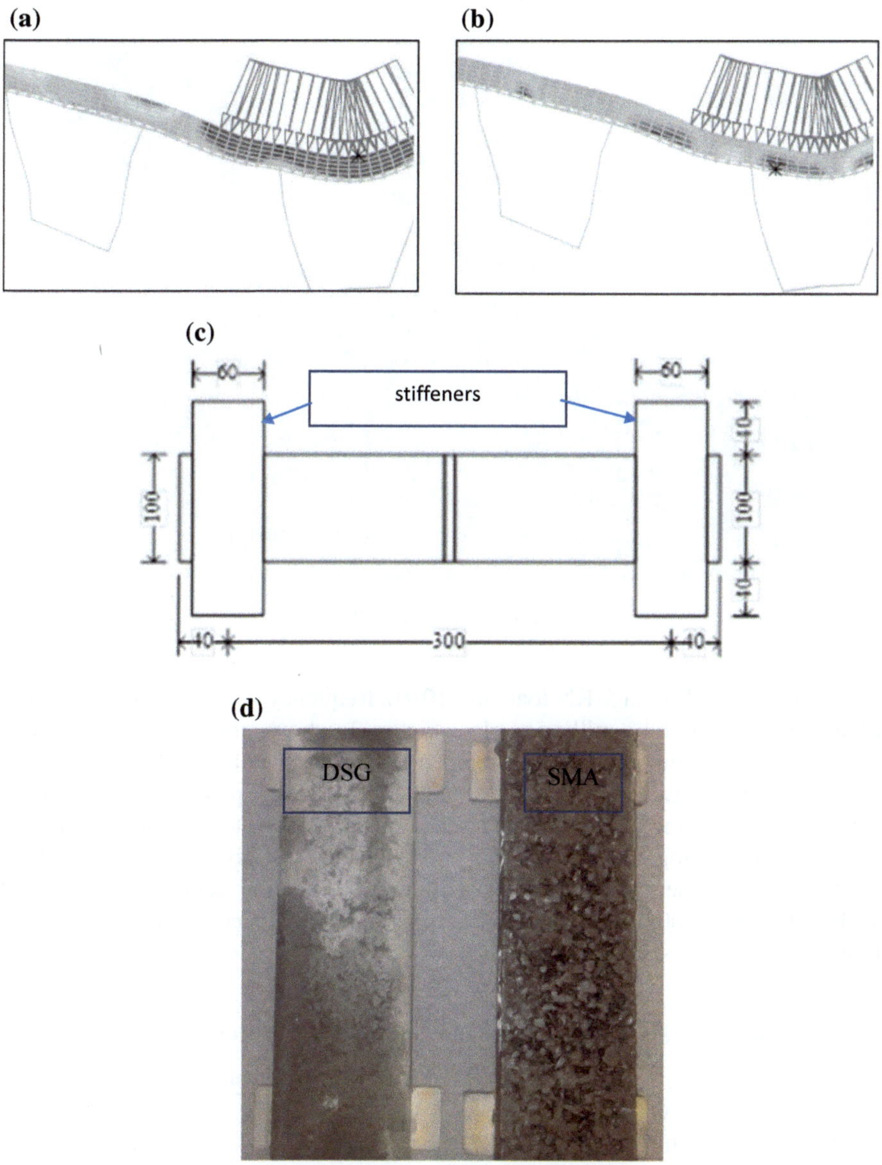

Fig. 3. Composite action of wearing surface and steel deck plate: **a** tensile and compressive stress distribution, **b** shear stress distribution **c** schematic of steel plate for testing, **d** completed composite beam specimen **e** beam flexural fatigue test equipment

the density of the beam is close to that of Marshall compacted samples. Figure 3d shows the completed composite beam specimens for both DSG and SMA epoxy asphalt mix as well as the surface texture for the two types of mix. A 3 KN load at the midpoint was given for the preconditioning and residual stress relieve purpose in the

(e)

Fig. 3. *(continued)*

first 50,000 cycles. Then a 5 KN load and 10 Hz frequency were used to test the beam until it fails or passes 12 million cycles. Figure 3e shows the composite beam test specimen and test equipment. Composite beam was tested in a temperature controlled chamber at 23 °C. Load force was applied to the steel plate directly, which is upside-down comparing to real vehicle load. Both the steel plate and epoxy asphalt mix deflect and recover together as elastically to withstand the given loading energy and also distribute the energy to adjust two stiffeners. Deflection values in Table 3 were recorded at around 300,000 cycles, by the time, deflection become fairly constant and before fatigue failure start to set in.

Table 3. Results for beam flexural fatigue test in full cured condition

	Type A binder	Type B binder	Type C binder
Dense graded mix			
Initial flexural stiffness (N/mm)	12,575.8	22,701.7	25,991.5
Final flexural stiffness (N/mm)	Failed	18,787.0	21,349.0
Fatigue performance (cycles)	2.0 million	12 million	12 million
Beam deflection (mm)	0.138	0.116	0.098
Stone mastic mixture			
Initial flexural stiffness (N/mm)	12,671.7	20,816.4	24,738.8
Final flexural stiffness (N/mm)	Failed	15,624.4	19,280.1
Fatigue performance (cycles)	0.8 million	12 million	12 million
Beam deflection (mm)	0.160	0.122	0.112

The factors of temperature and pavement thickness are not considered. The composite effect has been analyzed on two different steel plate thicknesses (11 and 16 mm) and pavement layer thicknesses is 40 mm.

The fatigue resistance seems to be related to stiffness, as can be seen in Table 3. The better performing of mixtures for fatigue resistance all have extremely high stiffness, which may come from the higher stiffness binder used (binder B and C) as well as the dense and stable mixture structure. Mixture is considered extremely good in fatigue resistance if the beam can pass 12 million cycles without reducing 50% of initial flexural stiffness. Final flexural stiffness was also recorded, as can be seen, none of them is reduced to 50% or less. Fatigue failure is defined as 0.5 mm midpoint deflection is reached and visible cracking can be seen often. Although Type A binder has good elasticity, this softer binder does not give enough fatigue resistance and suffered higher deflection. The similar phenomenon—stiffer epoxy asphalt mixture lead to lower steel deck deflection was also observed in other studies using various testing temperature, test frequency and wearing surface thickness (Seim and Ingham 2004; Yao et al. 2013). It can be seen that the deflection-to-span ratio ranges from 0.0004 to 0.0005 for Binder B and C beam samples. The value is lower than the 1/1200 (0.0008) deflection-to-span ratio specified in the new version of AASHTO bridge design specification (AASHTO 2010). This may indicates that it is practical to stiffen the steel deck and improve bridge pavement performance by using extreme high strength epoxy asphalt material.

The result—stiffer binder can provide better fatigue resistance is more or less contradictory to many previous studies that as the asphalt become oxidized and stiffer, the visco-elastic properties and fatigue resistance would greatly reduce (Islam and Tarefder 2015; Tang et al. 2015). This may be mainly due to the loss of adhesion and gain of rigidity after oxidation for conventional asphalt binder. To make it clear, the words "rigid" and "stiff" reflect different meaning and physical characteristics in this study. The cured epoxy asphalt binders with a 3 dimensional and continuous polymer structure become very stiff and greatly resistant to deformation; while aged asphalt binders become brittle and rigid, undergo no deformation and easier to crack under force.

Binder adhesion test was also conducted in this study. Direct measurement of binder and aggregate adhesion is difficult, mainly because of the aggregate's irregular shape and different sizes. Two types of pull-off test were done following a modified ASTM D 4541 method. In the first test, binder was sandwiched in between two pull-off steel tabs in 20 mm diameter (see Fig. 4a). In the second test, the pull-off tab was bonded on smooth cement mortar substrate, which is made of standard-graded Ottawa sand and Portland cement mix (Fig. 4b). To avoid neat epoxy asphalt binder flow out before hardening, samples were conditioned in a lower oven temperature at 60 °C for three days to reach full cured condition. Pull off tests were performed in room temperature (23 °C) and the test speed is 1.27 mm/min. Table 4 shows the pull-off bond strength for the three types of binders, and the pull-off bond strength is commonly referred to as "adhesion". Obviously, epoxy materials have the best adhesion to the steel as shown in Table 4. The binders also give good adhesion to cement mortar. The adhesion to cement mortar provides us an indirect indication of the binder bond strength to the aggregates as well. Studies in the University of Wisconsin-Madison

Fig. 4. Pull-off test samples for full cured epoxy asphalt: **a** steel to steel adhesion, **b** steel to cement mortar adhesion

Table 4. Pull-off bond strength test for epoxy asphalt in full cured condition

	Type A	Type B	Type C
Steel to steel (MPa)	4.88	8.77	9.55
Steel to cement mortar substrate (MPa)	1.75	5.19	6.03

show that, conventional asphalt binder and polymer-modified asphalt usually have an adhesive strength in between 1.5 and 2.5 MPa on limestone aggregates tested by Bituminous Bond Strength Test machine (Bahia et al. 2012), and the crumb rubber modified asphalt usually has an even lower adhesive strength (Bahia et al. 2012; Huang et al. 2016), as the crumb rubber is not an adhesive material and can hardly depolymerize homogenously in base asphalt.

If we compare the testing values from Tables 1 and 3 with Table 4, it can be easily noticed that, the adhesion measured by pull-off strength test seems to have a considerable influence and linear relationship to the modulus of epoxy asphalt binder as well as the stiffness of the beam sample. The correlation charts were made in Fig. 5. It is highly possible that the fatigue resistance and durability come from the adhesion properties of epoxy asphalt binder. However, the mechanism is not fully understood and more tests will be conducted to confirm the correlation in the future. In addition, all three tests, including binder tensile strength test, beam flexural fatigue test and pull-off bond test, were conducted in same temperature and should be comparable with no bias. Figure 6 shows the failure pattern after pull-off test. The figures further illustrate that Type A epoxy asphalt, like most conventional and polymer-modified asphalt, would have the adhesion failure mainly occur inside the thin bond layer, while the Type C binder have an adhesion failure totally from the aggregate and cement substrate.

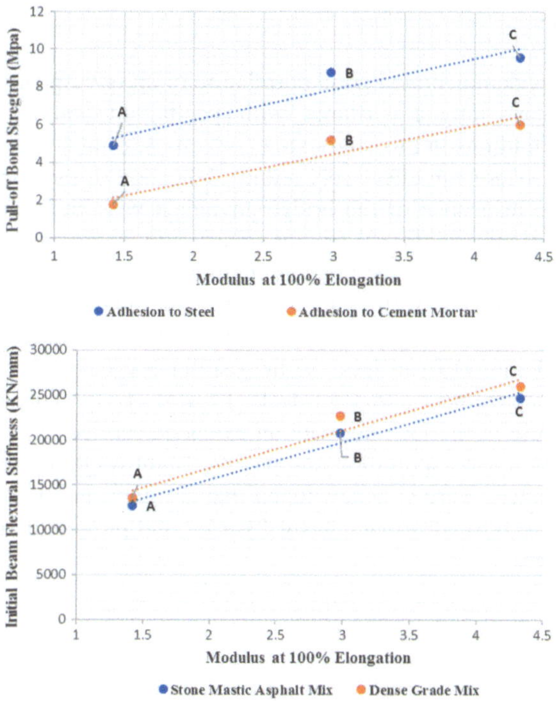

Fig. 5. Correlation among pull-off bond strength, binder modulus @ 100% elongation, and beam flexural stiffness

Fig. 6. Cement mortar substrate after pull-off test (left-Type A binder, middle-Type B binder, right-Type C binder)

4 Conclusions

Experimental results demonstrated good correlation of epoxy asphalt pavement on thin steel plates, particularly fatigue resistance, with specific physical bulk properties of the binder. Two major findings in this study are (1) Beam flexural fatigue tests show that a stiffer binder and mixture can effectively reduce deflection of the composite beam and improve the fatigue resistance of the bridge pavement without further increase pavement layer thickness and weight; (2) Binder adhesion to the aggregate matrix seems to play a pivotal role and the adhesive bond strength could have a close relationship with the beam flexural stiffness as well as the binder elastic modulus in low strain range.

The study also support that epoxy asphalt product can be used on extremely thin steel deck plate (11 mm or even less), to help stiffen the deck plate, mitigate fatigue problems on bridge pavement and steel deck. Type B epoxy asphalt seems to be the optimal binder for use in the job site, with an extended service time, which can give a longer time window for transportation and paving. Lightweight epoxy asphalt mixture with rich binder content were also developed that are good in fatigue resistance. This system can help reduce the dead weight by 13–15% and provide better skid resistance for bridge pavement.

References

Alabaster, D., Herrington, R., Water, J.C.: Long life open-graded porous asphalt. In: Third International Conference on Accelerated Pavement Testing. Spain (2008)

Alabaster, D., Herrington, P.R., Water, J.: Ultra long life low noise porous asphalt. J. Acoust. Soc. Am. (2012). https://doi.org/10.1121/1.4708028

American Association of State Highway Transportation Office—AASHTO: LRFD Bridge Design Specifications, 5th edn. American Association of State Highway and Transportation Officials, Washington, DC (2010)

American Institute of Steel Construction—AISC: Design Manual for Orthotropic Steel plate Deck Bridges. New York (1963)

Asphalt Institute: Mix design methods for asphalt concrete and other hot-mix types. MS-2, Lexington, KY (1979)

Bahia, H., Moraes, R., Velasquez, R.: The effect of bitumen stiffness on the adhesive strength measured by the bitumen bond strength test. In: 5th Eurasphalt and Eurobitume Congress. Istanbul (2012)

Federal Highway Adminstration—FHWA: Manual for design, construction, and maintenance of orthotropic steel deck bridges. Publication No. FHWA-IF-12-027 (2012)

Gaul, R.: Epoxy asphalt concrete—a polymer concrete with 25 years' expereince. Am. Concr. Inst. Publ. Symp. Publ. **166**(13), 233–251 (1996)

Huang, W.D., Quan, L.V., Xiao, F.P.: Investigation of using binder bond strength test to evaluate adhesion and self-healing properties of modified asphalt binders. Constr. Build. Mater. **113**, 49–56 (2016). https://doi.org/10.1016/j.conbuildmat.2016.03.047

Islam, M.R., Tarefder, A.R.: Study of asphalt aging through beam fatigue test. Transp. Res. Rec. **2505**, 115–120 (2015). https://doi.org/10.3141/2505-15

Jia, X.Y., Huang, B.S., Chen, S.J., Shi, D.W.: Comparative investigation into field performance of steel bridge deck asphalt overlay systems. Korean Soc. Civ. Eng. **20**(7), 2755–2764 (2016)

Lu, Q., Luo, S. Condition survey and analysis of first epoxy asphalt concrete pavement on orthotropic bridges in China—a ten year review report. Department of Civil and Environmental Engineering, University of South Florida, USA (2010)

Murakoshi, J., Yandori, N., Ishizawa, T., Toyama, N., Kosuge, T.: Study on effect of deck thickness of orthotropic steel deck on fatigue durability. J. Steel Constr. Eng. **19** (2012)

New Zealand Transport Agency—NZTA: SP/SM10:140905 Specification for dense graded and stone mastic asphalts (2014)

Organisation for Economic Co-operation and Development-OECD: Long-life surfacings for roads- field test results. http://dx.doi.org/10.1787/9789282108116-en (2017)

Seim, C., Ingham, T.: Influence of wearing surfacing on performance of orthotropic steel plate decks. Transp. Res. Rec. **1892**, 98–106 (2004). https://doi.org/10.3141/1892-11

Tang, S., Williams, R.C., Cascione, A.: Reconsideration of fatigue test for asphalt mixtures and binder containing high percentage RAP. Int. J. Pavement Eng. (2015). https://doi.org/10.1080/10298436.2015.1095895

Wu, J.P., Herrington, P.R., Alabaster, D.: Long-term durability of epoxy-modified open-graded porous asphalt wearing course. Int. J. Pavement Eng. (2017). https://doi.org/10.1080/10298436.2017.1366764

Yao, B., Wang, X., Hong, H., Cheng, C.: Research on the dynamic bending behavior of a composite beam model for the asphalt overlay on steel bridge deck. In: Transportation Research Board 92th Annual Meeting. Washington, DC (2013)

Changes in Chemical Fingerprints of Asphalt Binders Due to Aging

Shahriar Alam and Zahid Hossain(✉)

Arkansas State University, PO Box 1740, Jonesboro, AR 72467, USA
mdshahri.alam@smail.astate.edu, mhossain@astate.edu

Abstract. Asphalt binders remain exposed to aging from mixing and placement operations throughout their service life. The aging process is a chemical event, often known as oxidization, which hardens asphalt binders by causing changes to their chemical compositions. Stiffening also occurs when they are modified chemically to make them fit in certain environmental and loading conditions. Aging can accelerate stiffening of the modified binders because of the simultaneous presence of the modifier and oxidizing agents. This study evaluates the changes in the chemical fingerprints of asphalt binders in terms of their Saturates, Aromatics, Resins, and Asphaltenes (SARA) fractions and Fourier Transformation Infrared Spectroscopy (FTIR)—based functional groups. Two selected performance grade (PG) binders (PG 64-22) modified with Polyphosphoric Acid (PPA) and Styrene Butadiene Styrene (SBS) have been subjected to Rotational Thin Film Oven (RTFO) and Pressure Aging Vessel (PAV) aging. Asphalt binders from two different crude sources modified with different percentages of PPA, SBS and a combination of PPA and SBS were considered in this study. Due to the aging effects, the modified asphalt binders became abundant in solid phase than the unaged binders at a lower modification level, which in turn made the binders stiff. The changes in pattern were different for binders from two different crude origins. The findings from this study can help setting guidelines for level of chemical modification to asphalt binders.

1 Introduction

Asphalt binder is complex mixture of numerous hydrocarbons. The chemical complexity of asphalt binder is inherited from the maturation of the ancient living organisms, which is very susceptible to chemical oxidation by reaction with the atmospheric oxygen. The oxidative aging causes a mass embrittlement of asphalt binder, which is not desirable as far as the serviceability is the concern. Oxidation is also influenced a great deal by the mineral aggregates that stay together with the asphalt binder in the mixture and act like a mineral catalyst toward oxidation (Petersen 2009). Although, oxidation is not the sole cause of the mass embrittlement of the asphalt binder, it is still an area to be research on. The aging of asphalt binder happens when heat and air (i.e. oxygen) are present together. The pavement temperature is usually higher than that of the ambient temperature, since it absorbs all the heat around and thus becomes highly susceptible to oxidative aging. The sensitivity of asphalt binder to oxidative hardening

K. Zhang et al. (eds.), *Testing and Characterization of Asphalt Materials and Pavement Structures*, Sustainable Civil Infrastructures,
https://doi.org/10.1007/978-3-319-95789-0_5

is widely variable with the asphalt source and its chemical compositions (Petersen 2000). The Zaca-Wigmore experimental road test is regarded as the classic demonstration of this source dependence (Hveem et al. 1959). This aging, though happens during hot summer days, influences the low temperature susceptibilities of an asphalt binder. A pavement with less elasticity is vulnerable to traffic loadings. Sometimes, loading is not necessarily required for a crack to occur during the winter days.

An asphalt binder is often modified to achieve some rheological improvement both in the cases of low and high temperature seasons. According to the National Center for Asphalt Technology (NCAT), some of the specific reasons that justifies a modification of asphalt binder includes stiffening binders and mixtures to minimize rutting and softening the binders at low temperatures to improve the relaxation properties and strain tolerance (King 1999). With that motive, asphalt binder is often supplied to the highway agencies with chemical modification. Polyphosphoric Acid (PPA) and Styrene Butadiene Styrene (SBS) are some of the modifiers that have long been used in asphalt modification by refineries to achieve those above-mentioned superiorities. However, with all these modification asphalt binder is in fact pushed a little further toward the oxidative aging. The aging that might cause the hardening through molecular association in an unmodified binder can get a head start with the stiffening effect of these modifiers. It is very likely to expect the detrimental effects way before the termination of the life of a pavement. Therefore, the effect of aging on chemically modified asphalt binders are needed to be explored from an engineering point of view.

2 Literature Review

Asphalt binder is analyzed based on the family of like chemical compounds namely SARA fractions due to its inherent chemical complexity from the crude parent. The families are termed as Saturates (S), Aromatics (A), Resins (R), and Asphaltenes (A). All these compounds stay together as a mixture. The molecules in the complex mixture stay as agglomerates and bonded together by polar association forces such as hydrogen bonding and dipole moments (Petersen 2009). As temperature increases, this polar association forces gets weakened and broken and eventually increase the global hardening of the asphalt binder. The saturates fraction is a light straw colored oil, primarily hydrocarbon in nature and having a little aromaticity. The saturates fractions highly resistant to ambient air oxidation (Corbett and Merz 1975). The other three fractions remains at a trend of moving from more non-polar fractions to more polar fractions. The oxygen containing functional groups are formed due to this oxidation reaction. The amount of ketones that form due to oxidation is linearly related to the log viscosity of the asphalt binder (Lee and Huang 1973; Epps et al. 1986). Although the Asphaltenes fraction has been considered to be chemically inert by some researchers (e.g., Rostler and White 1959), the data presented in many literatures indicates that Asphaltenes are inherently extremely reactive with oxygen.

The aging of bitumen can be categorized into two groups. The main aging that happen due to oxidation is a irreversible process and ends up resulting in changes in the rheology of the binder. The other is termed as reversible one, which is the physical hardening (Bahia and Anderson 1993). The physical hardening may be attributed to molecular structuring which implies that the reorganisation of bitumen molecules (or bitumen microstructures) to approach an optimum thermodynamic state under a specific set of conditions (Branthhaver et al. 1993). The bitumen aging is a very complex process, which includes multiple variables as a aging contributor. The bitumen aging happens during the handling and storage operation as well as in field exposure for a longer period. In storage tanks, the asphalt is kept heated all the time so that it remains sufficiently fluid and ready to be used. The other variables that govern the aging are the mix, nature of aggregates, and even the particle size distribution, void content of the mix, production related factors, temperature and time. All these factors operate at the same time making the process of aging very complex (Lu and Isacsson 2002).

PPA and SBS have long been used in asphalt modification by refineries while some state departments of transportation (DOTs) have bans on PPA alleging it to be the cause of some premature pavement distresses such as striping. However, PPA and SBS have some beneficial implications too. PPA is an oligomer of H_3PO_4, which is obtained through dehydration of H_3PO_4 at high temperatures or by heating P_2O_5 dispersed in H_3PO_4 (Jameson 1959). SBS is an elastomer, which is frequently used in asphalt modification. SBS increases the kinematic and dynamic viscosity values of asphalt binders (Lu and Isacsson 1997). The aging induced the changes in the components of base binder and degradation of the SBS modifiers (Wu et al. 2009).

The aging is an oxidation phenomenon and this oxidation causes the formation of certain functional groups such as carbonyl and sulfoxides in asphalt binder (Lu and Isacsson 2002). The degree of formation of carbonyl and the sulfoxide functional groups can easily be identified using FTIR spectroscopy. Hossain et al. (2012) studied FTIR-based functional groups that developed due to warm-mix additive modification of. The FTIR gives signal with a peak at certain wavenumber values for certain functional groups.

3 Objectives

The study focused on finding the effects of chemical modification at macroscale. Asphalt binders from two different crude origins were analyzed for detecting the change in chemical composition at different aging conditions and also the functional groups that might occur upon both modification and aging. The objectives of this study were as follows: (i) perform short-term and long-term aging of the sample binders; (ii) measure the chemical composition of the both unaged and aged binders; (iii) detect any functional groups using IR spectroscopy; and (iv) analyze and interpret the data to observe any changes or pattern due to aging.

4 Materials and Methodology

Asphalt binder samples were collected from two approved suppliers of Arkansas Department of Transportation (ArDOT). They supply binders that fulfill the PG requirements as warranted by ArDOT. The unmodified and modified asphalt binder samples used in this study are listed in Table 1. The PPA used in modifying the asphalt binders in this study was of 105% grade. The SBS was commercially available as Vector Dexco 2411.

Table 1. Asphalt binder samples and the modification dosage

Sample name	Binder sources	Original binder + modification	True PG
S1B1	S1 (Canadian)	PG 64-22	68-32
S1B2	S1 (Canadian)	PG 64-22 + 0.25% PPA	70-30
S1B3	S1 (Canadian)	PG 64-22 + 0.50% PPA	72-36
S1Bex	S1 (Canadian)	PG 64-22 + 0.625% PPA	Not measured
S1B4	S1 (Canadian)	PG 64-22 + 0.75% PPA	74-40
S1B5	S1 (Canadian)	PG 64-22 + 2.00% SBS	78-31
S1B6	S1 (Canadian)	PG 64-22 + 0.5% PPA + 2.00% SBS	85-34
S2B1	S2 (Arabian)	PG 64-22	69-30
S2B2	S2 (Arabian)	PG 64-22 + 0.50% PPA	71-38
S2B3	S2 (Arabian)	PG 64-22 + 0.75% PPA	72-32
S2Bex	S2 (Arabian)	PG 64-22 + 0.875% PPA	Not measured
S2B4	S2 (Arabian)	PG 64-22 + 1.00% PPA	74-31
S2B5	S2 (Arabian)	PG 64-22 + 2.00% SBS	76-33
S2B6	S2 (Arabian)	PG 64-22+ 0.75% PPA + 2.00% SBS	81-32

4.1 SARA Analysis

The SARA analysis was intended for determining the percentages of certain families of chemical constituents in the tested asphalt binders. Any improvement in rheological properties happens through certain alteration of chemical constituents, which lead to a change in the percentages of chemical constituent fractions. The analysis was performed in accordance with "ASTM D 4124-09: Standard Test Method of Separating Asphalt into Four Fractions" (ASTM D4124-09 2009). The test specimen was put into reflux with n-Heptane for at least 3 h. To start a reflux, an asphalt specimen weighing 2.00 ± 0.30 g was taken in a round bottom flask. For each grams of asphalt specimen, 100 mL of n-Heptane (HPLC grade) was added to it. A stirring magnet was put into the flask. A Liebig condenser was fitted to the opening of that flask. The assembly was fastened with a clamp and set on a heating bath containing silicone oil or sand minus #20 US standard sieve. The heating bath was placed on hot plate and the temperature was set at 200 ± 50 °C and the stirring was set at 300 ± 50 rpm. This reflux operation caused the highly polar or the most solid fraction (i.e. the Asphaltenes) to precipitate. The n-Heptane dissolved the other three fractions except the Asphaltenes. Although the standard recommended to use iso-Octane, it was not capable enough to entirely dissolve the specimen. Therefore, n-Heptane was used to get the entire specimen dissolved. The other three fractions are collectively termed as Maltenes. The Maltenes were loaded onto a chromatographic column containing activated alumina (pH 9–10) of particle size 50–200 μm and allowed to elute under gravity. The Maltenes came out in a sequence as the Saturates, the Aromatics and the Resins. The Saturates fraction came first out of the activated alumina with n-Heptane. The napthene Aromatics fraction was eluted with consecutive application of toluene and toluene:methanol (50:50) solvents. A ultraviolet (UV) light of 366 nm wavelength was shined onto the column to monitor advancement of the Aromatics fraction. A fluorescent band was progressing down. After collecting all the fluorescent bands, the polar Aromatics or the Resins started to elute. It was collected lastly with trichloroethylene. The Saturates was colorless. The Aromatics was yellow or red in color while the polar aromatics or the Resins were black in color. All the eluted fractions were completely dried using a rotary evaporator and were reported as the percent fractions of the original sample. Sometimes a drying with chloroform was required to escape all the solvents out of the eluted fractions.

4.2 Fourier Transform Infrared Spectroscopy

The FTIR test is a spectroscopy technique applied on an asphalt binder to detect the presence or change in quantities of functional groups that might have occurred due to the modification (Yildirim 2007). In this test, a vibrational Infra-Red (IR) light is passed through the tested sample. When the natural vibrational frequency of a specific molecule matches the frequency of the IR radiation, the molecule absorbs the energy and increases the amplitude of the vibrational motion and detected as a peak in the interferogram. In this study, disposable Real Crystal IR cards were used for preparing the samples. The IR cards contained a KBr substrate. FTIR functions as a fingerprinting tool such as the case of humans. In FTIR, the natural vibrations of the covalent bonds among the molecules are exploited. Every type of bond has a certain natural frequency of vibration. Two of the same type of bond in two different compounds can exist in two slightly different molecular environments. Therefore, no two molecules of different structures have the same IR absorption pattern, which serves as the fingerprint for that specific compound (Pavia 2008). A Nicolet 8700 spectrometer was used in this study. The data acquisition and analysis was done using the Omnic 6.2 software.

5 Results and Discussion

The percentages SARA fractions for all the binder samples have been reported in Figs. 1 and 2 as stacked charts. Asphalt binder samples from the Canadian crude source (Source-1) were high in Asphaltenes content than those from the Arabian crude source (Source-2). Even though both binders were marketed as PG 64-22, the neat binder from Source-2 (S2B1) had an Asphaltenes content of 13.2% and the neat binder from Source-1 (S1B1) had an Asphaltenes content of 19.9%. On the other hand, S1B1 binder had a low polar fraction (Resins) compared to S2B1.

The Asphaltenes fractions seemed to serve as a viscosity building component, which resulted in an improved rutting resistance. Asphaltenes and Resins fractions are the solid and nearly solid fractions respectively in an asphalt colloidal structure. The overall strength is typically provided by the overall solid phase (Asphaltenes and Resins combined) of the binder. Therefore, despite having a lower Asphaltenes content, Source-2 binders were not poor in rutting resistance. Thus, despite having different amounts of solid species, both were marketed as PG 64-22 binders with nearly similar

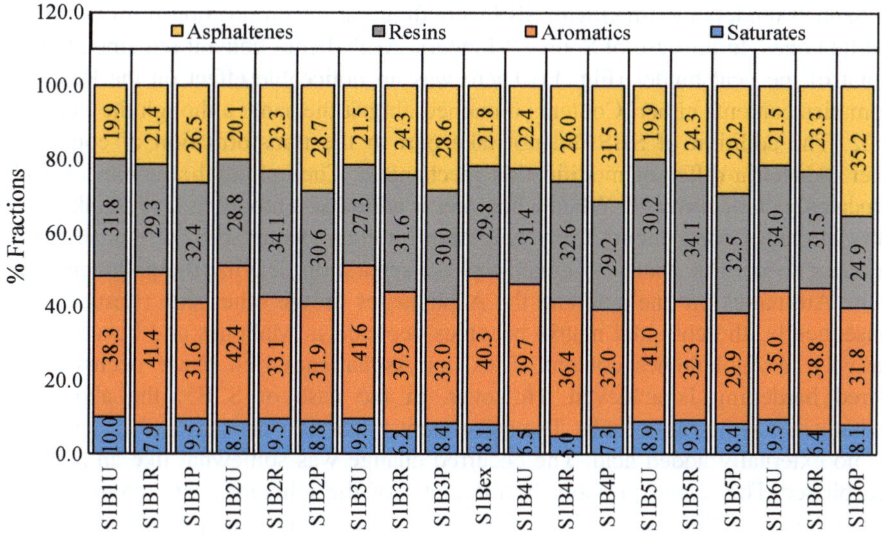

Fig. 1. SARA fractions of Source-1 asphalt binders at different aging conditions

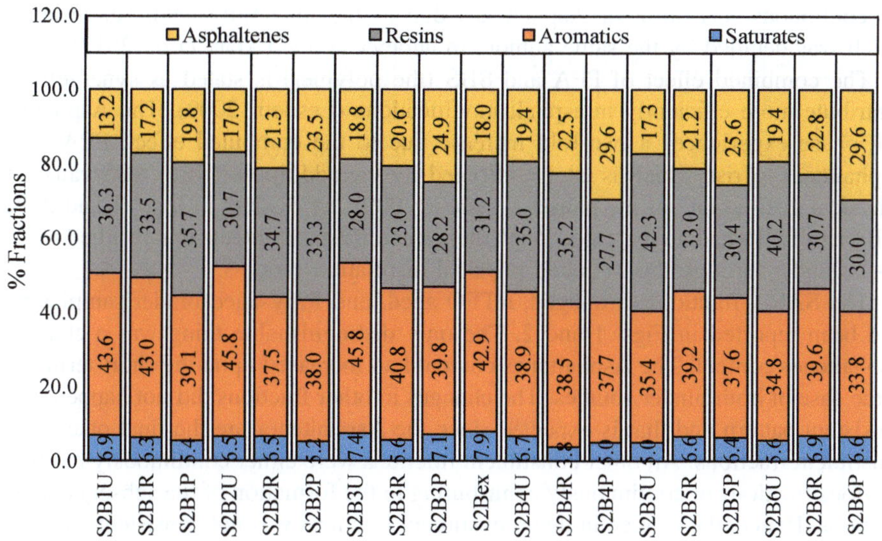

Fig. 2. SARA fractions of Source-2 asphalt binders at different aging conditions

rutting resistances. As per Robertson et al. (2001), an increase in polarity results in an increase in association and hence increases the stiffness. Source-2 samples had more polar Aromatics (i.e. Resins), which ensured the polar environment. The amount of PPA required to bump the PG from a neat binder could be observed in the test matrix (Table 1). A little higher amount of PPA was required for Source-2 binder to bump one PG grade than that from Source-1.

Asphalt binders modified with SBS were different in composition from that of PPA modified ones. In the case of Source-1 binders, Asphaltenes content remained the same as that of the neat binder (Fig. 1). There was no noticeable effect on the Resins and Aromatics contents either. Contents remained almost the same although the PG rose to 70-22. The addition of SBS did not take part in the transformation of compounds, rather induced a different modification mechanism. The colloidal structure of asphalt advocates such occurrence. When a polymer is added as a modifier, it is swollen by the light Aromatics components from the parent bitumen (Kraus 1982). Asphaltenes and polymers do not mix, but they create a phase separation, leaving the polymer swollen by the Aromatics on one side and the Asphaltenes on the other side (Lesueur 2009). Consequently, the colloidal matrix becomes depleted in Maltenes and hence enriched relatively in Asphaltenes content. Thus, an Asphaltenes rich phase occurs and the desired hardening is achieved. However, in the case of S2B5, the above-stated mechanism was not noticeable. The constituent fractions were changed although there was no externally added acid. The occurred change was somewhat like an acid modified binder. This could possibly be the effect of the inherent acidic environment of Source-2 asphalt binders, which promoted the association of compounds and made the conversion of Aromatics toward Asphaltenes easier. Although the neat binder from Source-2 contained a lower amount of Asphaltenes, a 2% addition (same as that of Source-1) of SBS resulted in a PG 70-22 binder. This occurrence could be attributed to the continuous formation of Asphaltenes due to the inherently acidic environment, which was detailed by the same authors in another article (Alam et al. 2017).

The combined effect of PPA and SBS (the polymer) is stated as synergetic and contribute more efficiently in asphalt modifications (Lesueur 2009). The Asphaltenes rich phase experiences a two-fold increase due to the combined effect. PPA causes Asphaltenes to rise which is later reinforced by the addition of SBS. SBS causes the "physical distillation" of the phases (Lesueur 2009). As evident in Figs. 1 and 2, both S1B6 and S2B6 had an increased Asphaltenes and Resins contents that resulted in a PG 76-22 binder through the so-called physical distillation process.

The SARA fractions of unaged, RTFO-aged, and PAV-aged binder samples have also been reported in Figs. 1 and 2. The only distinguishable thing was a change in Asphaltenes content. The Asphaltenes content had increased upon the short-term aging in the case of both binder sources. The changes in other fractions did not happen in any consistent pattern and that is expected since the Asphaltenes are the final phase of the constituent fractions. All other constituent fractions were either continuously fed by the previous fractions or continuously contributing to the formation of the subsequent polar fractions. Hence, the change in their constituent fractions was not consistent. However, there was a small but consistent change in the Saturates content. The Saturates content was reduced upon the RTFO-aging process. This phenomenon is analogous to the reduction in the sample mass in the RTFO-aging process. Upon PAV aging Asphaltenes content reached to a considerably higher percentage. The binders S1B6 and S2B6 were modified with the same amount of PPA as that of S1B3 and S2B3. In B6 binders from both sources, there was SBS in combination with PPA. It is claimed that SBS does not contribute to any sort of association process that might cause the formation of more Asphaltenes or polar species.

FTIR tests were conducted on both unaged and aged binder samples from both sources. From the FTIR spectra it was revealed that neither of PPA and SBS modification added any new functionality to the binder samples. The peak at certain wavenumbers displayed a higher signal, which implied that the aging introduced some increment in certain functional groups. Per Lamontagne et al. (2001), the change in the quantities of the functional groups could be obtained using the following equations.

Carbonyl Index (C=O),

$$I_{C=O} = \frac{\textit{Area of the Carbonyl band around } 1700\ \mathrm{cm}^{-1}}{\textit{Area of the spectral band between } 4000\ \mathrm{cm}^{-1} \textit{ and } 650\ \mathrm{cm}^{-1}} \tag{1}$$

Sulfoxide-Index (S=O),

$$I_{S=O} = \frac{\textit{Area of the Sulfoxide band around } 1030\ \mathrm{cm}^{-1}}{\textit{Area of the spectral band between } 4000\ \mathrm{cm}^{-1} \textit{ and } 650\ \mathrm{cm}^{-1}} \tag{2}$$

Tables 2 and 3 list the areas under the curves corresponding to the 1700 and 1030 cm^{-1} peaks and the corresponding indices based on Eqs. 1 and 2. The carbonyl index was the one that appeared in the spectra only after the PAV aging process. The oxygen molecules from the air got attached to the carbons upon aging. In the process of

Table 2. Different indices obtained from FTIR spectra for Source-1 binders

Samples	Aging condition	Total area (4000–650 cm^{-1})	Wavenumber (1700 cm^{-1})		Wavenumber (1030 cm^{-1})	
			Area	I (C=O)	Area	I (S=O)
S1B1	Unaged	1691.934	0.00	0.0000	11.33	0.0067
S1B2		1533.226	0.00	0.0000	7.58	0.0049
S1B3		1904.361	0.00	0.0000	10.06	0.0053
S1B4		1212.173	0.00	0.0000	0.02	0.0000
S1B5		1772.973	0.00	0.0000	0.82	0.0005
S1B6		952.508	0.00	0.0000	0.00	0.0000
S1B1	RTFO-aged	1713.094	0.00	0.0000	6.82	0.0040
S1B2		2091.482	0.00	0.0000	8.57	0.0041
S1B3		1791.154	0.00	0.0000	8.76	0.0049
S1B4		1429.215	0.00	0.0000	4.34	0.0030
S1B5		1954.239	0.00	0.0000	8.87	0.0045
S1B6		2267.321	0.00	0.0000	9.66	0.0043
S1B1	PAV-aged	1323.595	2.48	0.0019	7.24	0.0055
S1B2		1578.001	2.63	0.0017	9.86	0.0062
S1B3		2077.899	2.93	0.0014	15.14	0.0073
S1B4		1965.288	2.22	0.0011	24.25	0.0123
S1B5		2218.208	4.77	0.0022	16.08	0.0072
S1B6		1726.161	2.52	0.0015	7.71	0.0045

Table 3. Different indices obtained from FTIR spectra for Source-2 binders

Samples	Aging condition	Total area (4000–650 cm^{-1})	Wavenumber (1700 cm^{-1})		Wavenumber (1030 cm^{-1})	
			Area	I (C=O)	Area	I (S=O)
S2B1	Unaged	1656.524	0.00	0.0000	4.16	0.0025
S2B2		1470.115	0.00	0.0000	2.86	0.0019
S2B3		1811.484	0.00	0.0000	6.75	0.0037
S2B4		1817.391	0.00	0.0000	10.59	0.0058
S2B5		2013.758	0.00	0.0000	4.66	0.0023
S2B6		2149.632	0.00	0.0000	5.53	0.0026
S2B1	RTFO-aged	1723.880	0.00	0.0000	7.29	0.0042
S2B2		2010.876	0.00	0.0000	7.26	0.0036
S2B3		1205.605	0.00	0.0000	3.98	0.0033
S2B4		1470.358	0.00	0.0000	5.44	0.0037
S2B5		1660.154	0.00	0.0000	4.97	0.0030
S2B6		1775.001	0.00	0.0000	8.31	0.0047
S2B1	PAV-aged	1000.137	0.97	0.0010	7.16	0.0072
S2B2		1662.012	1.43	0.0009	7.56	0.0045
S2B3		1843.490	1.45	0.0008	10.92	0.0059
S2B4		1795.892	1.37	0.0008	13.08	0.0073
S2B5		1242.380	1.38	0.0011	11.43	0.0092
S2B6		1791.672	2.16	0.0012	7.86	0.0044

aging, the oxidation, dehydrogenation and crosslinking reactions occur at the same time (Siddiqui and Ali 1999). This occurrence leads to the appearance of more carbonyl (C=O) and sulfoxide (S=O) groups. Both the carbonyl and the sulfoxide indices increased upon the PAV aging. Like the Source-1 binders, both the carbonyl and sulfoxide indices increased upon aging. The carbonyl group was only visible for the PAV aged binders.

6 Conclusion and Recommendation

The Canadian crude binders were high in Asphaltenes fraction, whereas the Arabian Crude binders were much lower in the Asphaltenes content. However, both binders met the required Performance Grade (PG) specifications. The solid fraction alone does not contribute in the physical properties rather a combination of all the components implement changes in the physical properties of an asphalt binder. The required amount of PPA to cause a grade bump for Arabian crude source binders was higher than that of the Canadian crude source binders for similar grade bumps. It implies the influence of crude origin in setting up a binder modification methodology. The FTIR test data revealed that PPA modification did not introduce any new functional groups in the asphalt binder samples. Only the short-term and long-term aging caused the increase in certain functional groups especially the species that respond to oxidative aging. The recommended action for an asphalt binder to have the enhanced capability of sustaining severe aging could be making the binder deficient of polar species. This action could be performed either by taking out the highly polar fractions or adding the some less polar oil fractions like Aromatics from. This could prolong the saturation of the binder's colloidal structure and thus delay the aging.

References

Alam, S., Hossain, Z.: Changes in fractional compositions of PPA and SBS modified asphalt binders. J. Constr. Build. Mater. **152**, 386–393 (2017). http://dx.doi.org/10.1016/j.conbuildmat.2017.07.021

ASTM D4124-09.: Standard Test Method for Separation of Asphalt into Four Fractions. ASTM International, West Conshohocken, PA, USA (2009). https://doi.org/10.1520/d4124-09

Bahia, H.U., Anderson, D.A.: Glass transition behavior and physical hardening of asphalt binders (with discussion). J. Assoc. Asph. Paving Technol. **62** (1993)

Branthhaver, J.F., Petersen, J.C., Robertson, R.E., Duvall, J.J., Kim, S.S., Harnsberger, P.M., Scharbron, J.F.: Binder Characterization and Evaluation. Volume 2: chemistry (1993)

Corbett, L.W., Merz, R.E.: Asphalt binder hardening in the Michigan test road after 18 years of service. Trans. Res. Rec. (1975)

Epps, J., Petersen, J.C., Kennedy, T.W., Anderson, D., Haas, R.: Chemistry, rheology, and engineering properties of manganese-treated asphalts and asphalt mixtures. Trans. Res. Rec. (1986)

Hossain, Z., Lewis, S., Zaman, M., Buddhala, A., O'Rear, E.: Evaluation for warm-mix additive-modified asphalt binders using spectroscopy techniques. J. Mater. Civ. Eng. **25**(2), 149–159 (2012)

Hveem, F.N., Zube, E., Skog, J.: Progress report on the Zaca-Wigmore experimental asphalt test project. In: Symposium on Road and Paving Materials, ASTM International (1959)

Jameson, R.F.: The composition of the "strong" phosphoric acids. J. Chem. Soc. (Resumed), 752–759 (1959)

King, G.K.: Additives in asphalt. J. Assoc. Asph. Paving Technol. **68**, 32–69 (1999)

Kraus, G.: Modification of asphalts by block plymers of butadiene and styrene. Rubber Chem. Technol. **55**(5), 1389–1402 (1982). https://doi.org/10.5254/1.3535936

Lamontagne, J., Dumas, P., Mouillet, V., Kister, J.: Comparison by Fourier transform infrared (FTIR) spectroscopy of different ageing techniques: application to road bitumens. Fuel **80**(4), 483–488 (2001)

Lee, D.Y., Huang, R.J.: Weathering of asphalts as characterized by infrared multiple internal reflection spectra. Appl. Spectrosc. **27**(6), 435–440 (1973)

Lesueur, D.: The colloidal structure of bitumen: consequences on the rheology and on the mechanisms of bitumen modification. Adv. Colloid Interface Sci. **145**(1), 42–82 (2009)

Lu, X., Isacsson, U.: Influence of styrene-butadiene-styrene polymer modification on bitumen viscosity. Fuel **76**(14–15), 1353–1359 (1997)

Lu, X., Isacsson, U.: Effect of ageing on bitumen chemistry and rheology. Constr. Build. Mater. **16**(1), 15–22 (2002)

Petersen, J.C.: Chemical composition of asphalt as related to asphalt durability. Dev. Pet. Sci. **40**, 363–399 (2000)

Petersen, J.C.: A review of the fundamentals of asphalt oxidation: chemical, physicochemical, physical property, and durability relationships. Transportation Research E-Circular, Transportation Research Board (2009)

Rostler, F.S., White, R.M.: Influence of chemical composition of asphalts on performance, particularly durability. In: Symposium on Road and Paving Materials, ASTM International (1959)

Robertson, R.E., Branthhaver, J.F., Harnsberger, P.M., Peterson, J.C., Dorrence, S.M., McKay, J. F., Tauer, J.E.: Fundamental properties of asphalts and modified asphalts. Volume I: Interpretive report (No. FHWA-RD-99-212) (2001)

Siddiqui, M.N., Ali, M.F.: Investigation of chemical transformations by NMR and GPC during the laboratory aging of Arabian asphalt. Fuel **78**(12), 1407–1416 (1999)

Wu, S.P., Pang, L., Mo, L.T., Chen, Y.C., Zhu, G.J. Influence of aging on the evolution of structure, morphology and rheology of base and SBS modified bitumen. Constr. Build. Mater. **23**(2), 1005–1010 (2009)

Asphalt Mixtures that Dissipates Energy—Comparison of Conventional and Newly Developed Mixtures

Josef Zak[1(✉)], Jan Suda[1], Ondrej Dasek[2], and Petr Spacek[3]

[1] Faculty of Civil Engineering, CTU, Prague, Czech Republic
josef.zak@fsv.cvutcz
[2] Faculty of Civil Engineering, University of Technology, Brno, Czech Republic
[3] Division of Resources, Skanska a.s, Stockholm, Czech Republic

Abstract. Efforts to design cost-effective and high-quality pavements leads to the need to innovate the design of road construction materials. Part of this process is an innovation of procedures that are used to analyze the mechanical and physical properties of these materials. The paper deals with the description of results of ongoing project titled as Asphalt Mixtures Dissipating Energy. Thus, laboratory measured results of conventionally produced asphalt mixtures and selected asphalt binders are reported in the paper. Dynamic shear rheometer and bending beam rheometer was utilized to access linear viscoelastic properties of asphalt binders at low, intermediate and high temperatures. The susceptibility of asphalt mixtures to permanent deformation was evaluated by Hamburg Wheel Tracking Test and Uniaxial Shear Test. Four Point Bending Beam test was utilized to analyze fatigue resistance of asphalt mixtures. The paper summarizes principles used to develop new types of asphalt mixtures that, when loaded by traffic, better dissipate energy. The main target of the research project is to develop new asphalt mixtures that in situ prevent occurrence of rutting on the pavement. Thus, core part of the paper focuses on the analysis of linear viscoelastic properties related to pavement permanent deformation and its change depending on the change of input materials.

1 Introduction

The development of asphalt mixtures requires a multidiscipline solution: It is necessary to study the conditions that the product is exposed to (climatic influences, traffic loads) and as well as the mechanical properties of the material. The properties to consider should include the appropriate use of asphalt mixtures for energy dissipation in structures, the interaction with other pavement layers and pavement interaction with the subsoil. The solution is primarily determined by the material composition: The component proportions, material structure and the chemical bonds between the different components. Equally important is the development of a material that meets defined requirements on a macro-scale, so that it can be efficiently produced and processed in the volumes needed for infrastructure construction.

At present, national and European efforts are being made to innovate in the design of flexible pavements. That is, trying to find the most efficient pavement in terms of

K. Zhang et al. (eds.), *Testing and Characterization of Asphalt Materials and Pavement Structures*, Sustainable Civil Infrastructures,
https://doi.org/10.1007/978-3-319-95789-0_6

service life, functionality and whole-life cost; throughout design, production, construction and maintenance of the road. The need for innovation around construction materials has come from the increase in transport volume, the anticipated increase in wheel loads effects due to the increase in axle loads and new designs for heavy truck tires, and the effort to design economical and quality pavements. Part of this innovation is to review material characterization, to focus on the physical-mechanical nature of the processes and components that influence the behavior and overall value of the pavement as an asset.

Empirical methods of asphalt mix design can produce sub-optimal solutions by not fully utilizing the potential of the materials, and often they do not reflect the true behavior of the material in the pavement. These empirical approaches are gradually being replaced by a mechanical approach, but at the moment to only a limited extent. 'Dissipating asphalt' mixtures are a new material in this respect, developed through a new approach to the design and assessment of asphalt mixtures. The development is based on both conventional and innovative laboratory analyzes of the properties of individual components, additives, analyzes of their interaction, empirical and physical-mechanical properties.

This article presents the analysis of conventionally produced (reference) bituminous mixtures material properties. These analyzes are used in the next stages of the project for qualitative comparison. The key objective is to design of new asphalt mixtures that dissipate energy based on these data. This article also lists results from the design of asphalt mixes for wearing course.

2 Materials, Methodology and Testing

Analyzes were performed on three conventional mixtures for the surface courses—ACO 11+ 50/70, ACO 11+ PmB 45/85-65, SMA 11S PmB 45/80-65 and three conventional (reference) base courses—ACL 16S 50/70, ACL 22S 50/70 and ACL 22S PmB 25/55-65 (ČSN EN 13108-1 2008, ČSN EN 13108-5 2008).

Grading of the individual mixtures consists of crushed aggregates (granodiorite). The grading curve of these mixtures is shown in Fig. 1. Asphalt binder (50/70) and polymer-modified asphalt binder (PmB 45/80-65, PmB 25/55-65) were used for asphalt mixture production.

The assessment of the individual mixtures was carried out using two criteria: The first is the resistance to permanent deformation. This parameter has been determined by a Wheel Tracking Test on a small test device according to ČSN EN 12697-22 and a Uniaxial Shear Test according to (Zak et al. 2016). The second criterion was the fatigue parameters measured on the 4 PB-PR instrument according to ČSN EN 12697-24, but evaluated by the conventional method of 50% reduction of stiffness moduli according to ČSN EN 12697-24, Hopman and Pronk (Hopman 1989) and its modified method based on the relative reduction of the complex stiffness moduli—(Rowe) (Zak et al. 2013; Rowe 1996; Maggiore et al. 2014). These two methods are based on the same

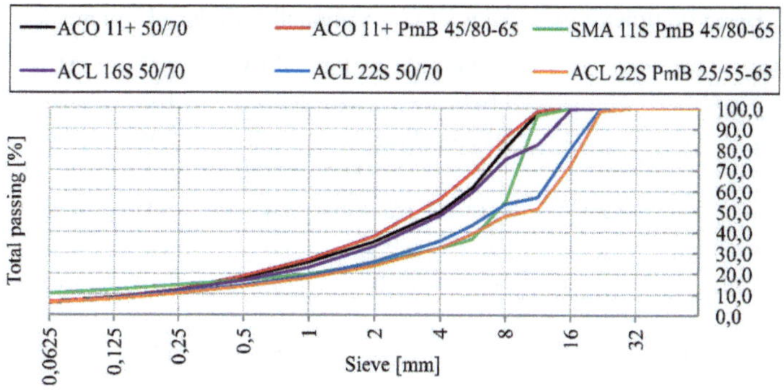

Fig. 1. Grading curves of reference asphalt mixtures

idea of the distribution of the fatigue test results in the controlled strain mode, in the form of the dissipated energy ratio into three phases and the definition of the cycle value (N1) at the phase II and III interfaces as resistance to fatigue (Zak et al. 2013).

The basic volumetric parameters, according to the relevant technical standards, were also determined for the produced mixtures:

• Maximum bulk density	according to ČSN EN 12697-5
• Compacted bulk density	according to ČSN EN 12697-6
• Air void contents	according to ČSN EN 12697-8
• Asphalt binder content	according to EN 12697-1
• Grading curve	according to ČSN EN 12697-2
• Voids in the mineral aggregate	according to ČSN 736160
• Voids filled with asphalt	according to ČSN 736160

Furthermore, the deformation parameters of asphalt mixtures were determined:

• Asphalt stiffness modulus IT-CY	according to ČSN EN 12697-26

All asphalt mixtures were produced at the asphalt plant.

3 Results and Discussion of Reference Mixtures

For the above-mentioned reference mixtures, the basic descriptive parameters (see Table 1) have been determined in accordance with the relevant standards. The stiffness modulus was determined by cyclic loading, using an indirect tensile loading test in controlled strain mode and at a defined loading pulse, on a Universal Testing Machine (UTM). For test mixtures, temperatures of 5, 15 and 30 °C were selected. This non-destructive test was performed on Marshall test specimens.

Table 1. Basic volumetric properties of conventionally produced (reference) mixtures

Asphalt mixture		ACO 11+	ACO 11 +	SMA 11S	ACL 16S	ACL 22S	ACL 22S
Asphalt binder		50/70	PmB 45/80-65	PmB 45/80-65	50/70	50/70	PmB 25/55-65
Air void	(%)	3.0	2.2	3.4	3.1	4.6	3.4
Asphalt binder content	(%)	5.6	5.9	6.6	5.3	4.3	4.2
Aggregate air void content	(%)	16.1	16.0	19.0	15.6	14.8	12.9
Asphalt binder content in the mixture	(% volume)	13.1	13.8	15.5	12.5	10.2	9.4
Voids filled with asphalt	(%)	81.1	86.2	81.9	80.0	68.8	73.2
Stiffness modulus—IT-CY (MPa)	5 °C	12,662	10,226	8373	18,208	23,722	16,012
	15 °C	6480	5514	4452	10,127	12,815	8862
	30 °C	1685	1808	1337	2721	3892	3028

Table 2. Parameters of resistance to permanent deformation

Asphalt mixture		ACO 11+	ACO 11+	SMA 11S	ACL 16S	ACL 22S	ACL 22S
Test temperature		60 °C	60 °C	60 °C	50 °C	50 °C	50 °C
Asphalt binder		50/70	PmB 45/80-65	PmB 45/80-65	50/70	50/70	PmB 25/55-65
Wheel Tracking Test	WTS_{AIR} (%)	0.135	0.034	0.013	0.017	0.014	0.020
	PRD_{AIR} (%)	5.85	3.10	1.90	1.00	1.50	1.30

Resistance to permanent deformation was tested on an air-conditioned Hamburg Wheel Tracking test—small test device. This teste method parameters examine the susceptibility of the asphalt mixture to a permanent deformation, demonstrated by the depth of the rut caused by the repeated travel of the load wheel at a defined temperature. Asphalt mixtures for the permanent deformation resistance test were compacted in the roller compactor according to (ČSN EN 12697-33+A1 2007). The degree of compaction of all samples was maintained in the range of 99.0–101.0%. The test was carried out at 60 °C for the surface layers and at 50 °C for the base layers. The results are shown in Table 2 and Fig. 2.

Other recorded parameters describing the durability of asphalt mixtures against permanent deformations were obtained from the Uniaxial Shear Test (UST) (Zak et al. 2016). The recorded parameters were; the shear modulus, the coefficients of regression accumulated permanent deformation, the number of cycles to permanent shear strain, the constant shear strain values at 5000 and 10,000 cycles, and the increment of permanent strain (see Table 3; Fig. 3). The test was carried out at 60 °C for the surface

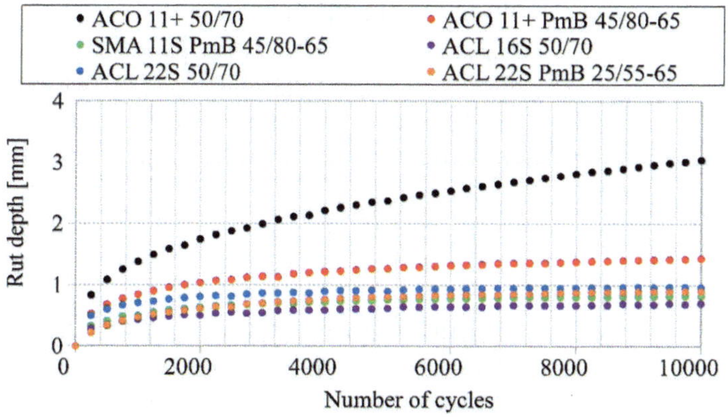

Fig. 2. Permanent deformation during Wheel Tracking Test

layers and at 50 °C for the base layers. The test specimens were manufactured on a gyrator according to ČSN EN 12697-31 in the compaction range of 99.0–101.0%.

The results of the Wheel Tracking Test indicate that ACL 16S, SMA 11S and ACL 22S show higher resistance to permanent deformations. This higher resistance is mainly due to the suitable aggregate composition and the use of a modified binder to fill the air voids to the optimum extent. However, it is difficult to derive the causal relationship between the effects on resistance to permanent deformation from the test parameters, mainly due to the low-resolution capability of the Wheel Tracking test. On the other hand, the ACO 11+ mixture tends to produce permanent deformations, however, the use of a modified binder reduces this deficiency. For ACO 11+ , the hard workability has to be noted.

The resistance to permanent deformation was further monitored using the UST. The UST results indicate a significant effect of the aggregate grain curve and the bituminous binder on the resulting resistance to permanent deformations. The UST allows a better differentiation of individual mixtures from several perspectives: shear modulus, increment of permanent deformation, ratio of elastic and plastic deformation. This makes it possible to identify the susceptibility of asphalt mixtures to permanent deformations.

In the UST test, the effect of the aggregate composition and the binder used is more evident. It can be seen from the evaluation of asphalt mixtures used in surface layers. These mixes exhibit a higher resistance to permanent deformation, especially the SMA 11S mixture and of the ACL 22S mixture with polymer-modified binder. Figure 4 shows test specimens of the ACL 16S mixture, which exhibits the lowest resistance to permanent deformation. The picture was taken after testing so that the accumulated permanent deformation is evident.

The resistance of asphalt mixtures to degradation processes is an important material property because it affects the life of the entire highway construction.

The effect of a parameter characterizing fatigue resistance is apparent from the established design methodologies of pavement structures: It is usually found as an exponent in the calculation and thus exponentially affects the value of the calculated

Table 3. Shear parameters

Shear parameters	ACO 11+ 50/70		ACO 11+ PmB 45/80-65		SMA 11S PmB 45/80-65	
	50 °C	60 °C	50 °C	60 °C	50 °C	60 °C
Shear modulus @ 1000c (MPa)	6.9E+01	5.5E+01	5.1E+01	4.5E+01	5.8E+01	6.0E+01
Regression of accumulated shear strain [–]						
parameter A	4.5E−03	7.1E−03	5.0E−03	7.4E−03	5.0E−03	1.6E−03
parameter B	1.6E−01	1.5E−01	1.3E−01	1.3E−01	1.8E−01	2.8E−01
Number of cycles to [–]						
1% γ	2.0E+08	9.2E+05	5.7E+08	3.8E+06	6.1E+05	7.1E+05
3% γ	2.0E+08	9.2E+05	5.7E+08	3.8E+06	6.1E+05	7.1E+05
5% γ	2.6E+10	3.3E+07	1.2E+11	2.8E+08	6.8E+06	1.3E+07
Permanent shear strain						
at 5000 cycles (mγ)	1.8E+01	2.2E+01	1.5E+01	2.1E+01	1.8E+01	1.7E+01
at 10,000 cycles (mγ)	2.0E+01	2.8E+01	1.6E+01	2.2E+01	1.9E+01	2.2E+01
Increment of permanent deformation (mγ/10^3)	4.4E−01	1.1E+00	2.1E−01	3.1E−01	2.6E−01	1.1E+00
Shear parameters	ACL 16S 50/70		ACL 22S 50/70		ACL 22S PmB 25/55-65	
	50 °C	60 °C	50 °C	60 °C	50 °C	60 °C
Shear modulus @ 1000c (MPa)	7.6E+01	5.1E+01	6.8E+01	4.9E+01	6.2E+01	5.6E+01
Regression of accumulated shear strain [–]						
parameter A	1.2E−02	4.1E−03	1.6E−02	5.2E−03	5.4E−03	7.8E−03
parameter B	1.7E−01	2.8E−01	9.3E−02	1.5E−01	1.5E−01	1.1E−01
Number of cycles to [–]						
1% γ	9.3E+04	3.1E+03	8.2E+02	1.5E+03	1.9E+05	6.6E+09
3% γ	9.3E+04	3.1E+03	8.2E+02	1.5E+03	1.9E+05	6.6E+09
5% γ	6.4E+05	2.1E+04	2.2E+05	3.3E+04	1.1E+07	4.3E+11
Permanent shear strain						
at 5000 cycles (mγ)	3.2E+01	3.7E+01	3.5E+01	4.2E+01	1.9E+01	1.9E+01
at 10,000 cycles (mγ)	3.5E+01	5.0E+01	3.6E+01	4.7E+01	2.1E+01	2.0E+01
Increment of permanent deformation (mγ/10^3)	6.3E−01	2.8E+00	2.9E−01	1.1E+00	3.3E−01	2.7E−01

pavement resistance (TP 170 2004; Epps et al. 2002; Monismith 2012; Balay, Caron, a Lerat, b.r.). The fatigue parameter value is therefore essential both for the design and for the life of the road construction.

Fatigue is defined as a consequence of a repeated load on the internal structure of the compacted asphalt mixture. It is manifested by the gradual decrease of the complex stiffness moduli in relation to the number of load cycles. The life of the asphalt mixture, which is defined as the number of load repetitions to sample failure, can therefore be measured. In a controlled strain test, the damage of the bond between the stressed

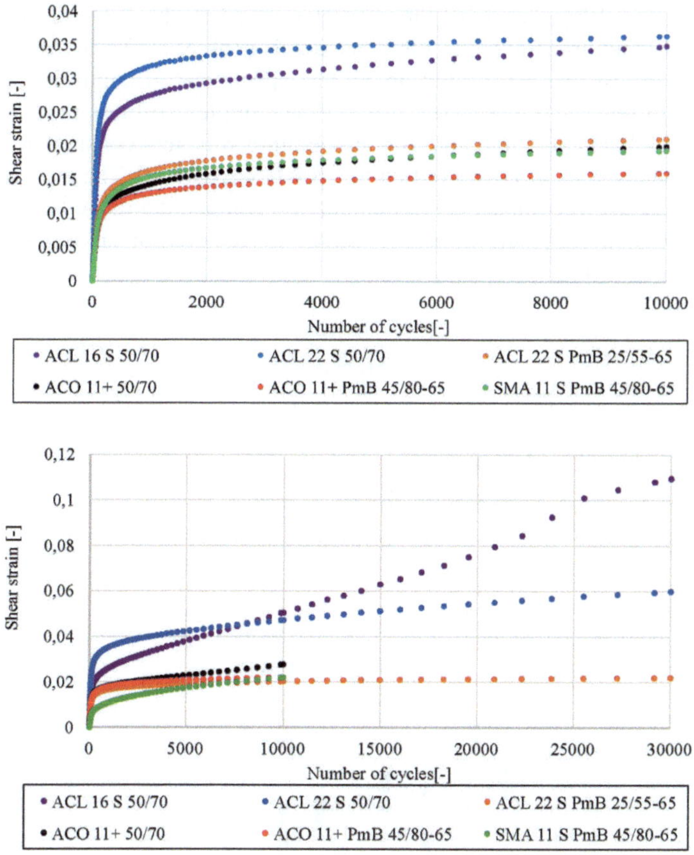

Fig. 3. Permanent deformation during Uniaxial Shear Test

Fig. 4. Accumulated permanent deformation (ACL 16S 50/70)—Uniaxial Shear Test (UST)—50/60 °C

particles is recorded by a drop in force resistance, when the force resistance of the damaged bond is not transmitted to the other bonds (Zak et al. 2013; Zak et al. 2015).

Each fatigue test was performed on 18 test specimens in controlled strain mode with three strain levels, so that a 50% drop in the complex stiffness modulus occurred in the 10^4–2.10^6 load cycle interval. The temperature was kept constant at 20 ± 1 °C. The load frequency was 30 Hz according to ČSN EN 13108-20. The test samples were produced on a segment compaction equipment at a compaction range of 99.0–101.0%. The slabs were then cut to the defined dimensions required for the test samples. Test results are shown in Fig. 5.

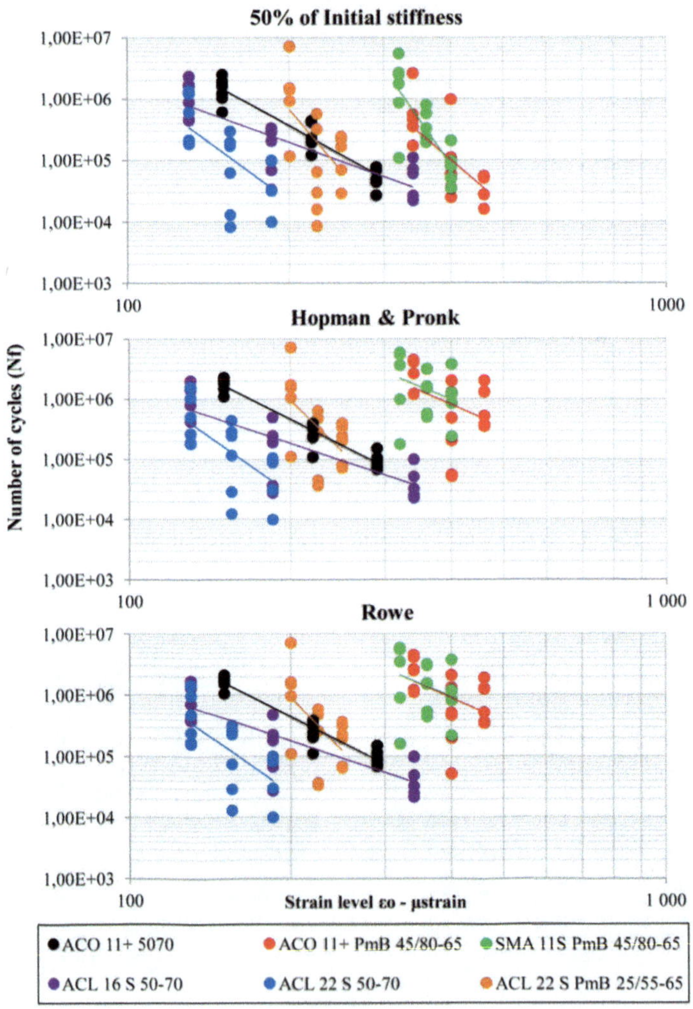

Fig. 5. Fatigue resistance

It follows from the results that the highest values of fatigue resistance are shown in SMA 11S for the surface layers and the ACL 22S mixture for base layers. Both of these mixtures contain a relatively high amount of a polymer modified binder. Figure 6 shows a comparison between the different mixtures, for the conventional 50% drop in stiffness modulus and energy dissipation methods (Rowe, Hopman and Pronk), by evaluating the fatigue parameter e_6 derived from the fatigue curve (Zak et al. 2014).

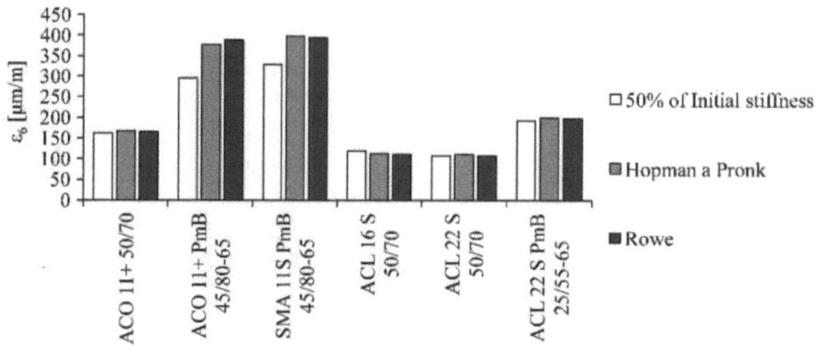

Fig. 6. Strain ε_6 derived from the Wöhler curve at 10^6 cycles

These test results prove that the resistance to fatigue of asphalt mixtures determined by the energy dissipation method, is higher than the 50% method of stiffness reduction. The results also show that this is more evident in asphalt mixtures with modified asphalt binders, which implies that the asphalt mixtures with polymer modified binders better dissipate energy in terms of fatigue resistance. The fatigue properties were determined based on the procedure described in (Zak et al. 2015; Zak et al. 2013).

4 Design and Evaluation of Asphalt Mixtures that Dissipates Energy—SMA 16

The basic principle of this design, methods and principles is to develop an asphalt mixture that, to the greatest extent, resists the accumulation of permanent deformations and at the same time exhibits the longest fatigue life.

For the development of the dissipating asphalt mixtures, five asphalt binders available on the market were selected and evaluated. Selected binders were evaluated primarily with respect to their mechanical properties. Individual binders were evaluated using both empirical and mechanical tests, with emphasis on changes caused during the aging of the asphalt binder. Short-term aging has been simulated to describe the changes that occur during the production, laying and compacting of the asphalt mixture, and long-term aging. From this research according to the tested parameters a modified binder PmB 45/80-85 was selected.

The choice of asphalt mixtures results from the evaluation of the reference mixtures or from the above mentioned assessment of the functional parameters of the reference mixtures. The study of the reference mixtures implies that tested variants with higher maximum grain size better resists to permanent deformations. Based on this, a mixture of SMA with a maximum grain size D = 16 mm was chosen. Thus the proposed new mixture that dissipates energy is SMA 16. Five variants of the gradation curves (see Fig. 7) were created to match the grading curves according to ČSN EN 13108-5. The aggregate used for new asphalt mixtures comes from the same source as the aggregate used in reference mixtures.

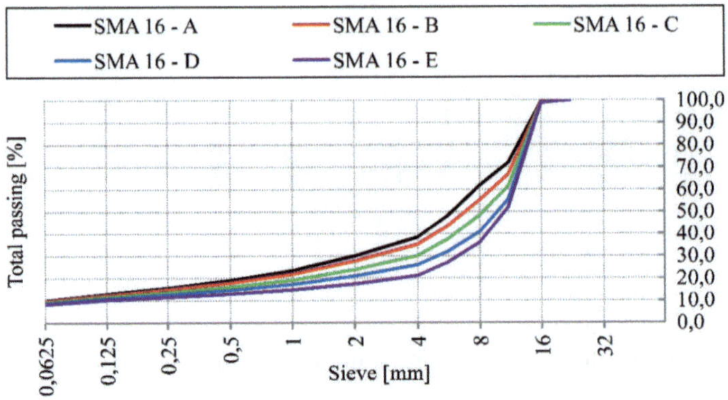

Fig. 7. Asphalt mixture grading curves—SMA 16

Table 4. Volumetric properties—SMA 16—new asphalt mixtures

SMA 16					
Variant	Binder content (%)	Air voids (%)	Voids in mineral aggregate (%)	Asphalt binder content (% volume)	Voids filled with asphalt (%)
A	5.3	3.1	15.6	12.5	80.1
	5.7	2.6	16.0	13.4	83.7
	6.2	1.4	16.1	14.6	91.0
B	5.3	3.1	15.5	12.4	80.1
	5.7	2.5	15.9	13.4	84.5
	6.2	1.9	16.5	14.6	88.2
C	5.3	3.5	15.9	12.4	78.0
	5.7	3.3	16.6	13.3	80.0
	6.2	3.1	17.6	14.5	82.3
D	5.3	4.4	16.7	12.3	73.7
	5.7	3.7	16.9	13.2	78.3
	6.3	2.9	17.5	14.7	83.6
E	5.3	7.8	19.7	11.9	60.3
	5.7	7.3	20.1	12.8	63.6
	6.3	6.3	20.2	13.9	68.9

For individual variants (A-E) of the grading curves an optimization was performed based on the graduated quantity (5.3–5.7–6.2 wt%) of the added asphalt binder. For these sub-variants, the basic volumetric parameters are presented in Table 4.

Individual variants (A-D) were subsequently optimized to meet the requirements of ČSN EN 13108-5. The variant E did not fulfill the standard requirements so it was excluded from further testing. The next step was to determine the resistance to permanent deformation in Hamburg Wheel Tracking Test and the IT-CY stiffness modules as first indicators of mixtures suitability for further testing. The results are shown in Table 5.

Table 5. Optimized variants—SMA 16

SMA 16					
Variant		A	B	C	D
Binder content (% weight)		5.4	5.5	**5.7**	**5.8**
Air voids (%)		2.8	2.9	**3.3**	**3.4**
Voids in mineral aggregate (%)		16.7	15.8	**16.5**	**16.9**
Asphalt binder content (% volume)		14.0	12.9	13.2	13.5
Voids filled with asphalt (%)		83.0	81.6	79.9	79.8
Wheel Tracking Test	WTS_{AIR} (%)	0.039	0.025	0.021	0.016
	PRD_{AIR} (%)	2.2	2.2	**1.5**	**1.1**
Stiffness modulus—IT-CY (MPa)	5 °C	13,935	15,146	**14,331**	**14,927**
	15 °C	5939	7023	**6905**	**7457**
	30 °C	1244	1507	**1571**	**1736**

Table 6. Shear parameters of SMA 16 C, D

Shear parameters	SMA 16-C		SMA 16-D	
	50 °C	60 °C	50 °C	60 °C
Shear modulus @ 1000c (MPa)	6.34E+01	6.40E+01	6.65E+04	5.15E+04
Regression of accumulated shear strain [–]				
parameter A	1.09E−02	9.78E−03	4.93E−03	9.29E−03
parameter B	9.81E−02	9.54E−02	2.39E−01	5.84E−02
Number of cycles to [–]				
1% γ	3.79E+04	2.29E+04	1.37E+02	4.91E+01
3% γ	3.79E+04	2.33E+07	2.77E+03	1.35E+07
5% γ	3.50E+06	5.89E+08	1.85E+04	1.30E+10
Permanent shear strain				
at 5000 cycles (mγ)	1.75E+01	1.40E+01	9.24E+00	7.86E+00
at 10,000 cycles (mγ)	1.82E+01	1.49E+01	9.65E+00	8.23E+00
at 30,000 cycles (mγ)	–	2.08E+01	–	1.77E+01
Increment of permanent deformation (mγ/10^3)	1.38E−01	1.69E−01	8.15E−02	7.33E−02

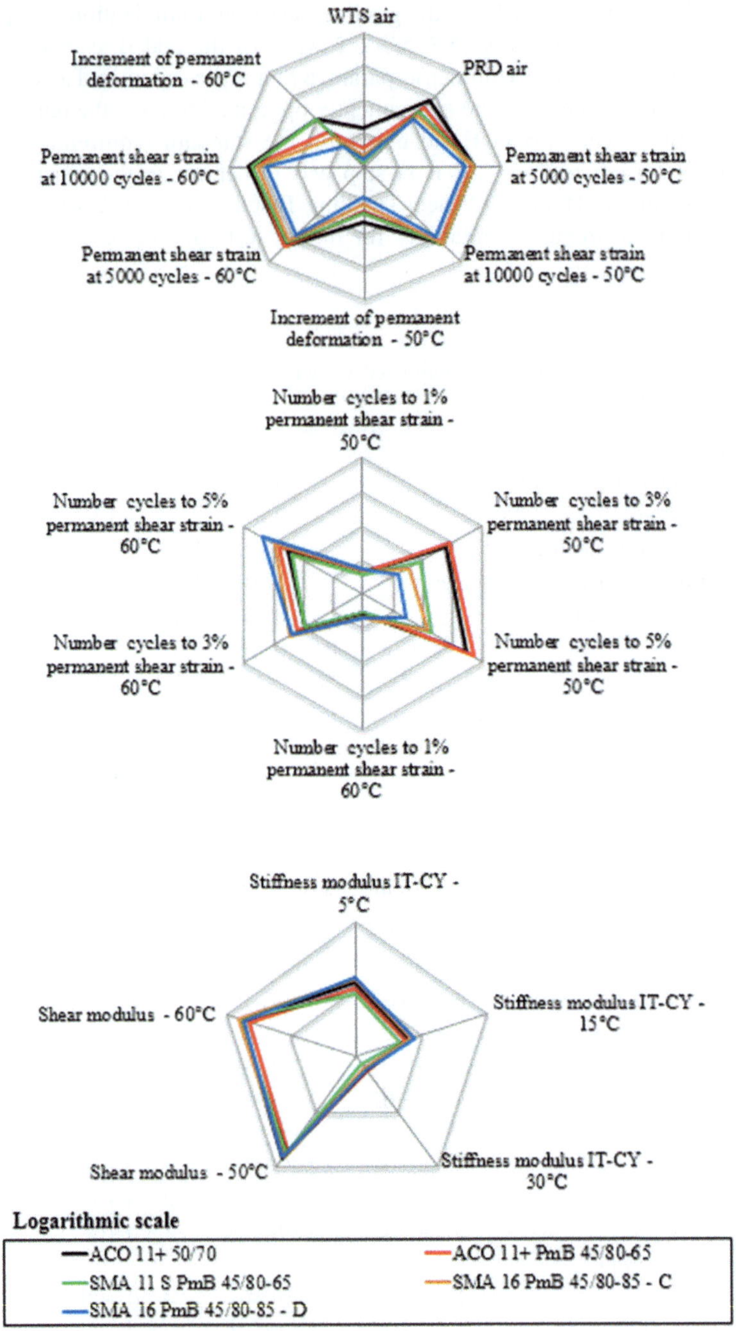

Fig. 8. Schematic comparison of the test mixtures

The results show a clear influence of the aggregate composition (grading curves), respectively with the increase in the content of coarser grains in the mixture the resistance to the permanent deformation increases. The influence of binder content is also significant. From the results obtained in Table 5, variants C and D were further evaluated and analyzed in Uniaxial Shear Test. Results are shown in Table 6.

The SMA 16-C and SMA 16-D variants are compared to the reference mixtures and results are presented in Fig. 8. Here it is quite clear that the SMA 16-D mixture qualitatively shows a very good resistance to permanent deformations compared to the other reference mixtures under consideration. This method of designing the mixture using empirical and functional tests made it possible to design a qualitatively more efficient asphalt mixture in terms of selected parameters. In addition the Uniaxial Shear Test is capable to better distinct than Hamburg Wheel Tracking between asphalt mixtures resistance to permanent deformation.

5 Conclusions

The objective of the research project is not to compare conventionally produced asphalt mixtures between each other, but to examine the influence of the individual components of the mixtures on the resulting mechanical and physical properties. Then, on the basis of this knowledge, the objective was to propose new mixtures that dissipate energy more effectively. From all studied new variants of asphalt mixtures the variant D has a better performance and asphalt mixture properties are reported in this paper.

It is not assumed that a newly designed asphalt mixture will be cheaper than conventional materials, but will be a better value material with better functional properties, thus reducing the cost of repairs, maintenance and reconstruction over the service life of the infrastructure. The economic benefits from the use of these materials will be predominantly felt by administrators and owners of transport infrastructure.

This article recommends the possibility of maximizing the useful properties of bituminous mixtures by using more accurate evaluation procedures and by using more accurate test procedures with higher resolution capabilities such as a Uniaxial Shear Tester.

The presented test results imply that the asphalt mixtures with polymer modified binders better dissipate energy in terms of both permanent deformation and fatigue resistance.

Acknowledgements. This paper was elaborated within the research project No. FV10526 in the program TRIO of the Ministry of Industry and Trade of the Czech Republic.

References

Balay, J.M., Caron, C., Lerat, P.: Adaptation of the French Rational Road Design Procedure to Airfield Pavement: The Alize-Airfield Software. Atlantic City, New Jersey, USA

ČSN EN 12697-33+A1: Bituminous mixtures—test methods for hot mix asphalt—part 33: specimen prepared by roller compactor. Czech Office for Standards, Metrology and Testing (2007)

ČSN EN 13108-1: Bituminous mixtures—material specification—part 1: asphalt concrete. Czech Office for Standards, Metrology and Testing (2008)

ČSN EN 13108-5: Bituminous mixtures—material specification—part 5: stone mix asphalt. Czech Office for Standards, Metrology and Testing (2008)

Epps, J.A., Seeds, S., Schulz, T., Alavi, S.H., Ashmore, C., Monismith, C.L., Deacon, J.A., Harvey, J.T., Leahy, R.B.: Recommended Performance-Related Specification for Hot-Mix Asphalt Construction: Results of the WesTrack Project. Transportation Research Board—National Research Council, Washington, D.C. (2002). http://onlinepubs.trb.org/onlinepubs/nchrp/nchrp_rpt_455a.pdf

Monismith, C.L.: Flexible pavement analysis and design-a half-century of achievement. In: Geotechnical Engineering State of the Art and Practice, pp. 187–220. American Society of Civil Engineers (2012)

TP 170.: Navrhování vozovek pozemních komunikací. Ministerstvo dopravy České republiky (2004). http://www.pjpk.cz/TP%20170.pdf

Zak, J., Stastna, J., Zanzotto, L., MacLeod, D.: Laboratory testing of paving mixes—dynamic material functions and wheel tracking tests. Int. J. Pavement Res. Technol. **3**(6), 147–154 (2013)

Zak, J., Monismith, C.L., Jarušková, D.: Consideration of fatigue resistance tests variability in pavement design methodology. Int. J. Pavement Eng. **2014**(8), 1–6 (2014)

Study on Moisture Susceptibility of Foamed Warm Mix Asphalt

Yingbiao Wu[1,2], Qingyi Xiao[3], Jinjin Shi[1(✉)], and Peiliang Zhang[1]

[1] Cangzhou Municipal Engineering Company Limited, Cangzhou, China
czszjsk@126.com
[2] Hebei Province Road Materials and Technology Engineering Technology Research Center, Cangzhou, China
[3] School of Civil Engineering, Civil Engineering Research Center of Hebei Province, Hebei University of Technology, Tianjin, China
chaphd@sina.com

Abstract. Warm mix asphalt (WMA) is the asphalt concrete mixtures that are produced at lower temperatures than the hot mix asphalt. There are important environmental and health benefits associated with the reduced production temperatures and this can also potentially improve pavement performance. However, there exists one concern about the moisture damage of WMA because of the foaming water addition. The objective of the paper is to evaluate the moisture susceptibility of laboratory foamed WMA under moisture conditioning procedures and its laboratory-performance. Nuclear magnetic resonance (NMR) measurements was employed to evaluate the residual moisture content in the foamed asphalt. Two moisture conditioning procedures were used, including immersion Marshall test procedure and Lottman test freeze–thaw procedure. The moisture-conditioned specimens were tested using the marshall stability test, the indirect tension (IDT) tests, and the Hamburg wheel tracking test. Stripping Inflection Point (SIP) of Hamburg rutting test gave an indication of the time when the mixture begins to exhibit moisture damage. Both HMA's and Formed WMA's SIP were greater than 20,000 cycles, which indicated that the foamed WMA had almost same water resistance as HMA.Through comprehensive analysis, foaming water did not seem to have significant effects on moisture susceptibility.

1 Introduction

Hot mix asphalt (HMA) is an excellent paving material with over 100 years of proven performance. It has been recorded that 90% hard-surfaced roads in the world is HMA [1, 2]. The HMA is produced at an elevated temperature in either batch or drum mixing plants and then compacted at temperatures ranging from 140 to 170 °C [3, 4]. In the past few decades, advances in asphalt technology such as polymer modified asphalt and the use of angular aggregate improved the anti-rutting performance (stone matrix asphalt, for example). The emphasis on the compaction for quality assurance have resulted in further increases in HMA mixing temperatures up to a limit of 175 °C. These high temperatures are linked to the increased emissions and fumes from HMA plants [5]. In addition, the production process of HMA consumes considerable energy

K. Zhang et al. (eds.), *Testing and Characterization of Asphalt Materials and Pavement Structures*, Sustainable Civil Infrastructures,
https://doi.org/10.1007/978-3-319-95789-0_7

to dry the aggregate and heat all materials prior to mixing and compaction. With concerns about global warming and fossil energy consumption, the asphalt industry is always seeking to lower its carbon emission. Warm Mix Asphalt (WMA) is a step in that direction. The use of warm mix asphalt (WMA) technology results in reduced production and paving temperatures without sacrificing the quality of the final product. This has led to a wider ranges of available production temperatures that may be employed by the contractor.

There have been a number of products and processes introduced to the marketplace to produce WMA over the last several years [6]. Among them, binder foaming becomes the most favorable choice for most contractors to produce WMA. The steam then foams the liquid binder, increasing its volume and allowing it to coat the aggregate at a relative lower temperature than that of traditional HMA. In current mechanical plant foaming processes in the United States, cold water is injected into a hot binder varying from 140 to 165 °C. The cold water turns to steam when it contacts with the hot binder, which will create an increased volume in binder. The amount of water used in producing the foam is normally varied from 1.0 to 3.0% by weight of binder. Foaming improves workability at lower temperatures in two ways: (1) it increases the volume with foaming, which easier to coat aggregate particles, and (2) it reduces mixture viscosity through shear thinning, which makes the mix more workable [7]. This allows the reduction mixing and placing temperatures, but it may introduce a new set of conditions that are not readily accounted for during the selection of materials and mixture design.

Several technical concerns also arise about WMA. The foamed WMA has trace moisture inside so that the major issue of moisture-induced damage remain to be answered [8, 9]. The lower temperatures can result in undrying of the aggregate and trapping water in the coated aggregate, which could lead to the moisture damage. Adequate moisture damage analysis needs to be tested to ensure that WMA will perform equivalently with HMA. Lottman test (AASHTO T283 Asphalt Mixtures Standard Method Test for Moisture Damage) is predominately used to determine HMA moisture damage susceptibility. However, current knowledge on diffusion and evaporation of moisture in foamed WMA is limited, and Lottman test may not be an accurate method. The water susceptibility of known mixtures may not satisfactorily match the observed behavior of the mixtures for a number of data groups.

2 Research Objective and Scope

The objective of the paper is to evaluate the moisture susceptibility of laboratory-foamed WMA under moisture conditioning procedures and the laboratory performance of WMA. The WMA was produced in a commonly-used foaming method in the China. The WMA and HMA were produced based on mixture used in a pavement project in Hebei province. The paper employed the nuclear magnetic resonance (NMR) measurements to evaluate the residual moisture content in the foamed asphalt. Two moisture conditioning procedures were used, including immersion Marshall test procedure and Lottman test freeze–thaw procedure. The moisture-conditioned specimens were tested using the marshall stability test, the indirect tension (IDT) tests, and the Hamburg wheel tracking test. The TROXLER™ Hamburg test was conducted for 20,000 cycles.

3 Materials and Specimen Preparation

3.1 Materials

The coarse aggregate in the mixture consisted of limestone rock, 10–15 mm stone, 5–10 mm stone, 3–5 mm stone screenings and machine-made sand with a nominal maximum aggregate size (NMAS) of 13.2 mm. The test indexes were shown in Tables 1, 2, 3. Asphalt binder used in the mixture was A-70 (Table 4). Based on Marshall mixture design method, optimum asphalt content (OAC) was 4.58%, and its mechanical indexes and volumetric indexes were shown Table 5. Lab scale foamed asphalt equipment was Wirtgen™ WLB10, and foaming conditions were 150 °C asphalt and 1% foaming water by asphalt weight. WMA mixture was designed following HMA mixture design. The compaction temperature was determined at which WMA had almost same value of volumetric indexes. The moisture susceptibility of WMA and HMA of a base mixture (called "AC-13" mixture in China) for a pavement project in Hebei province was evaluated and compared (Table 1).

Table 1. Coarse aggregate test results

Test item	Unit	Test result			Technical requirement
		(10–15) mm aggregate	(5–10) mm aggregate	(3–5) mm aggregate	
Crush value	%	15.2	–	–	≤ 26
Abrasion value (Los Angeles)	%	19.3	13.8	10.8	≤ 28
Apparent density	g/cm^3	2.818	2.750	2.746	≥ 2.6
Water absorption	%	0.76	0.73	0.72	≤ 2.0
<0.075 mm particle content	%	0.3	0.6	0.8	≤ 1

Table 2. Fine aggregate test results

Test item	Unit	Test results	Technical requirement
Apparent density	g/cm^3	2.827	≥ 2.5
Mud content	%	1.7	≤ 3.0
Sand equivalent	%	84.6	≥ 60

Table 3. Filler test results

Test results		Unit	Test results	Technical requirement
Apparent density		g/cm^3	2.764	$\geq$2.5
Moisture content		%	0.4	$\leq$1
Hydrophilic coefficient		–	0.82	$\leq$1
Particles size range	<0.6 mm	%	100	100
	<0.15 mm	%	90.3	90–100
	<0.075 mm	%	82.1	75–100
Appearance		–	Non-caking	Non-caking

Table 4. 70$^{\#}$ origin asphalt technical indexes

Test item	Test results	Technical requirement
Penetration (25 °C, 5 s, 100 g)/0.1 mm	64.5	60–80
Ductility (15 °C, 5 cm/min)/cm	>100	$\geq$100
Soft point (°C)	46.4	$\geq$46

Table 5. Mechanical indexes and volumetric indexes of HMA AC-13 with OAC

Asphalt content (%)	Bulk specific density	VV (%)	VMA (%)	VFA (%)	Stability (kN)	Flow (0.1 mm)
4.58	2.474	4.584	14.110	67.595	11.25	16.7
Specification requirement	–	4–6	$\geq$14	65–75	$\geq$8	15–40

3.2 Sample Preparation

In order to determine residual moisture amount of foamed asphalt after foaming and mixing, this study employed the NMR measurements by using the nuclear magnetic resonance (NMR) proton spin relaxation time (T2) distribution to predict foamed asphalt based on A2 peak area [10]. Micro MR40 NMR instrument in the study has 0.53T magnet strength, 23 MHz proton resonance frequency and keeps at a constant temperature of 32 °C. Because of the limitation of test space, about 10 g binder was poured into glass ampulla with plastic cap, the inside diameter is 20 mm and height 95 mm. Two types of binder were tested including foamed asphalt and foamed asphalt after RTFOT at 120 °C.

All the loose WMA and HMA mixtures were mixed by lab mixer. The WMA samples were compacted in lab using Marshall compaction device and Superave gyratory compactor. For immersion Marshall test and Lottman test, cylindrical samples were compacted to a height of 63.5 mm and diameter of 101.6 mm. For Hamburg wheel tracking tests, cylindrical samples were compacted to a height of 150 mm and a diameter of 150 mm. The target air voids for all mixture samples were controlled at

7 ± 0.5%. Two moisture conditioning procedures, immersion Marshall test procedure and Lottman test procedure, were conducted in the study [11, 12].

4 Results and Discussion

4.1 Moisture Amount of WMA by the NMR

A proton can be viewed as a micro magnet, since it bears a slight magnetic moment originating from its spin. In a porous material within a fixed external magnetic field, all protons will tend to align along with the direction of the magnetic field, resulting in magnetization. The magnitude of magnetization is proportional to the number of protons and inversely proportional to the temperature according to the Curie law. By applying a pulse of radio, the magnetization can be temporarily oriented to an unstable direction. After the pulse is removed, each proton spin emits the absorbed energy and returns to the initial equilibrium position through a series of relaxation processes which can be characterized by using a NMR instrument [13]. During the returning process, the voltage induced by the magnetization is detected, which decreases as the magnet returns to its stable position. The voltage decay curve as described above is also called the free-induction decay (FID) curve.

The magnitude of T_2 depends on the binding effect of pore water and the pore geometry. In general, the bound water or the water in small pores relaxes faster than the free water or the water in large pores. As a general rule, the relaxation time T_2 is less than 60 ms for the pore water in micropores or tightly bound water. If the T_2 ranges from 60 to 300 ms, this is for the water in micropores or loosely bound water. And if T_2 is larger than 300 ms, this is for the water in macropores or bulk water. For convenience, the area under the T_2 distribution curve is termed as the "peak area". The peak area represents the population of water molecules and serves as an indicator of the amount of unfrozen water content in the soil sample. So, the area value is satisfied the following equation [13]:

$$\mathrm{M(t)} = \sum_i A_i \exp\left(\frac{-t}{t_{2i}}\right) \tag{1}$$

where A_i is i component percentage, t_{2i} is relaxation time of i component.

After the NMR measurements, the measured data are analyzed by using an inversion procedure based on the Fourier transformation, as shown at Fig. 1. There are two peaks in the curve. One peak is near $T_2 = 0.1$ ms and another peak is near $T_2 = 5$ ms. T_2 signal intensity has a good relationship with binding effect of pore water and the pore geometry. When T_2 is less than 1 ms, it means substance in pores has very high viscosity or like solid. Based on carbon and hydrogen compound NMR data, the left peak is 1H in binder NMR signal curve, whereas the right peak is water signal curve. A2 peak area represents the amount of water content for the base binder sample which is very low as compare to A1.

Figure 1 shows the distribution curves of the relaxation time T_2 foamed binder with different foaming consumption after RTFOT at 120 °C. Through analysis of curves, it

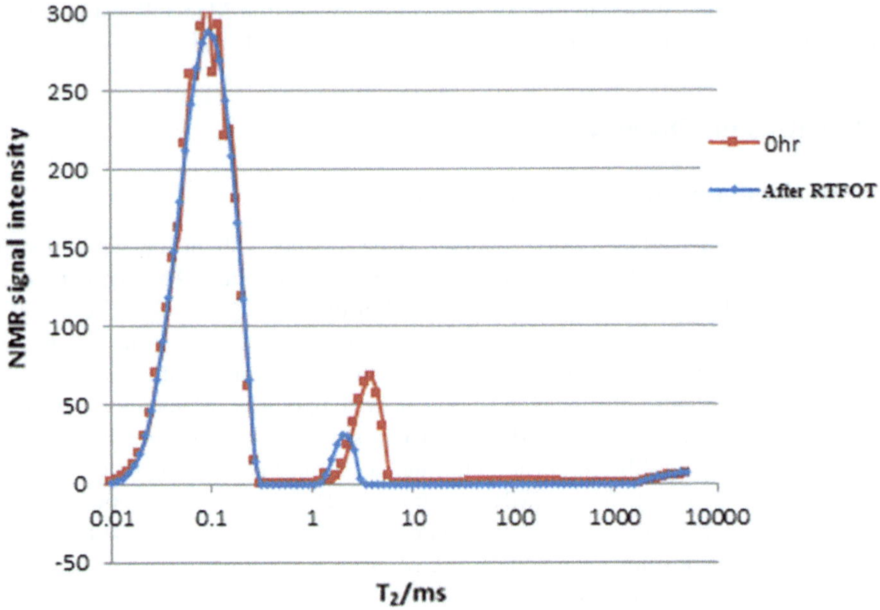

Fig. 1. T_2 distribution curve for foamed asphalt (1% foaming water)

is easily found that signal curve A2 peak weakens and curve area decreases with increasing curing time, whereas those A1 remain relatively unchanged. According to the relationship between moisture content and NMR signal intensity, signal changes shows that some water in foamed binder evaporated by curing conditioning. Analysis results of A2 peak area and moisture content after RTFOT at 120 °C are listed in Table 6. The results showed that there was a little residual water keeping in WMA, which meant approximately 50% foaming water evaporates during mixing and placing processes.

Table 6. Results of A2 peak area

Initial moisture content	Item	0 h	RTFOT at 120 °C
1%	A2 peak area	31.144 (RSD* 7.6%)	17.870 (RSD 3.6%)
	Moisture content (%)	1.0	0.57

*Relative standard deviation is abbreviated in RSD

4.2 Immersion Marshall Test

Moisture damage was also an important premature damage, which decreased pavement service level and reduced pavement life. Moisture typically reduces stiffness of the binder and/or mastic through moisture diffusion, and degrades the adhesive bonding between the binder/mastic and aggregate particles. Therefore, a loss of mixture internal strength results in premature distresses such as rutting, raveling, and fatigue cracking. At present, immersion Marshall test was one of two tests to evaluate water stability of mixture. The test moisture conditioning was performed by pre-soaking the specimens at 60 °C in hot water for 48 h. Results were shown in Table 7.

Table 7. Immersion Marshall test results

Mixture type	Compaction temperature (°C)	VV (%)	Marshall stability (kN)	Marshall stability after 48 immersion (kN)	Retained Marshall ratio (%)
HMA	150	4.584	11.25	10.94	97.22
WMA	120	4.611	9.06	8.83	97.46
Specification requirement	–				≥ 80

4.3 Lottman Test

Lottman test procedure was conducted in the study. At present, Lottman test was the other test to evaluate water stability of mixture. The moisture conditioning according to AASHTO T283 was performed by pre-soaking the specimens to the desired degree of water saturation and then finishing one cycle of freeze–thaw. Results were shown in Table 8.

Table 8. Lottman test results

Mixture type	Compaction temperature (°C)	VV (%)	Splitting strength (MPa)	Splitting strength after freeze-thaw (MPa)	Retained strength ratio (%)
HMA	150	4.584	1.50	1.34	89.33
WMA	120	4.611	1.16	1.01	87.35
Specification requirement	–				≥ 75

Through analysis of immersion Marshall test and Lottman test results, water stability of WMA met specification requirement, and almost is equal to that of HMA. According to immersion Marshall test and Lottman test, asphalt foaming process had no effect on water stability. But some studies have shown that both tests may not be accurate tests. The water susceptibility of known mixtures may not match the observed behavior of the mixtures for a number of data groups. An alternative test method used was Hamburg wheel tracking devices. Hamburg wheel tracking devices was used in this study to compensate traditional test [14].

4.4 Hamburg Wheel Tracking Test

TROXLER™ Hamburg wheel tracking tester was employed. The test produces damage by rolling a steel wheel across the surface of two SGC samples that are submerged in hot water at 50 °C. The samples were loaded until the maximum number of cycle (20,000) was reached. The stripping inflection point was determined from the graph of rut depths versus number of cycles. This defines the number of passes at which moisture damage starts adversely affecting the mixture. The higher the stripping inflection point, the less the asphalt mixture is likely to strip or be damaged by moisture [15].

The results from TROXLER™ wheel tracking test are shown in Fig. 2 and Table 9. The results of WMA mixture and the HMA mixture showed no distinctive stripping damage after 20,000 cycles. The WMA and HMA mixtures did not show any inflection points. With foamed asphalt, the rut depth of the WMA samples did not significantly decreased because of trace residual moisture, which indicates that the WMA mixtures would perform well in terms of moisture susceptibility.

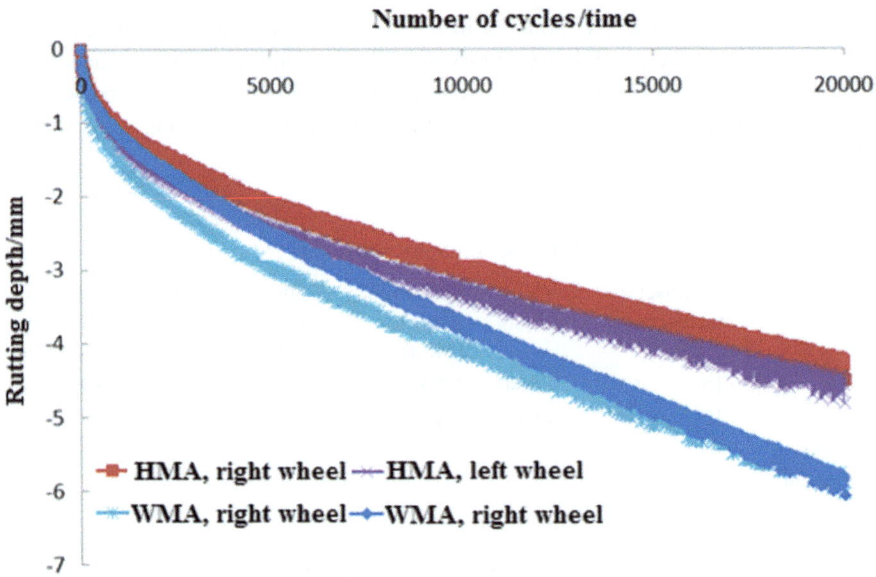

Fig. 2. Hamburg wheel tracking test results

Table 9. Hamburg wheel tracking test results

Mixture type	Compaction temperature (°C)	VV (%)	Bulk density (g cm^{-3})	Rutting depth (mm)
HMA	150	4.44	2.478	Left wheel: 4.82 Right wheel: 4.51
WMA	120	4.55	2.475	Left wheel: 6.08 Right wheel: 6.04

5 Conclusion and Summary

In order to comprehensively analyze the effect of residual foaming water on WMA moisture susceptibility, this study employed the NMR measurements to evaluate the residual moisture contents in foamed asphalt. The immersion Marshall test and Lottman test were conducted to determine moisture damage resistance, and the Hamburg wheel tracking test was conducted to investigate mixture stability in hot and humid environment. The conclusions are as follows.

According to the relationship between A2 peak areas and moisture contents in NMR theories, the NMR test results showed there was a very little trace residual moisture entrapped in WMA. About 50% foaming water was residual in mix, others evaporated during mixing and placing processes.

Results of traditional tests for moisture susceptibility showed that asphalt foaming process had no effect on moisture stability.

Wheel tracking test results showed the WMA mixture and the HMA mixture had almost the same stability in hot and humid environment. With foamed asphalt, the rut depth of the WMA samples was a little larger than that of HMA, but did not significantly decreased because of trace residual moisture.

The above findings were drawn from the results of the laboratory tests. A long-term field performance of the moisture damage of the WMA test sections will be evaluated in the future.

References

1. National Asphalt Pavement Association (NAPA): Warm mix asphalt: the wave of the future NAPA (2010). Accessed 15 July 2010
2. Wu, Y., Guo, Y., Shi, J., Zhang, B.: Application of warm mix technology based on foamed asphalt in road construction. Municipal Engineering Technology, China **4**(29), 17–21 (2011)
3. Kuennen, T.: Warm mixes are a hot topic. Better Roads **74**(6). James Informational Media, Des Plaines, IA (2004)
4. Yang, Y., Wu, Y., Qiao, J.: Experimental study on aspha-min warm-mix binder performance. Chinese and foreign road, China **3**(34), 291–294 (2014)
5. Stroup-Gardiner, M., Lange, C.R., Carter, A.: Quantifcation of emission potential from binders using mass loss and opacity measurements. Int. J. Pavement Eng. **6**(3), 191–200 (2005)
6. Prowell, B.D., Hurley, G.C.: Warm-mix asphalt best practices. Quality Improvement Series 125. National Asphalt Pavement Association, Lanham, MD (2007)
7. Fort, J.-P., Graham, G., Farnham, J., Sinn, S., Rowe, G.M.: I-55/I-57 Warm mix project in Missouri contractor's WMA experience. In: Proceedings of 2nd International Conference on warm mix asphalt. National Asphalt Pavement Association, Lanham, MD. www.asphaltpavement.org (2011). Accessed 22 Oct 2011
8. Federal Highway Administration (FHWA): Warm mix asphalt (2012). Accessed 19 Jan 2012
9. Hurley, G., Prowell, B.: Evaluation of potential processes for warm mix asphalt. J. Assoc. Asphalt Paving Technol. **75**, 41–90 (2006)
10. Wu, Y., Shi, J., Xiao, Q., Yu, Z.: 1H-NMR Relaxation and application study of evaporable water in foamed asphalt. In: 4th Geo-China International Conference, pp. 163–171

11. Highway research institute of the transportation department: Test methods of aggregate for highway engineering(JTG E42-2005), pp. 8–133. China Communications Press, Beijing (2005)
12. Highway research institute of the transportation department: Technical specifications for construction of highway asphalt pavements(JTG F40-2004), pp. 23–39. China Communications Press, Beijing (2004)
13. Ishizaki, T., Maruyama, M., Furukawa, Y., Dash, J.G.: Premelting of ice in porous silicaglass. J. Cryst. Growth **163**, 455–460 (1996)
14. Bhusal, S., Cross, S.A.: A laboratory study of warm mix asphalt for moisture damage potential and performances issues. Master thesis. Oklahoma State University (2008)
15. Izzo, R., Tahmoressi, M.: Use of the Hamburg wheel-tracking device for evaluating moisture susceptibility of hot-mix asphalt. Transport Res. Rec. **1681**, 76–85 (1999)

Effect of Different Adding Methods of Nano-TiO_2 on Photocatalytic Degradation of Automobile Exhaust on Asphalt Mixture

Qingyi Xiao[1(✉)], Xiangyang Chen[1], Yun Yang[2], and Jinguo Zhao[2]

[1] School of Civil and Transportation, Hebei University of Technology, Tianjin, China
chaphd@sina.com
[2] Tianjin Yongyang Highway Engineering Group Co., Ltd., Tianjin, China

Abstract. As to photocatalytic degradation of automobile exhaust by TiO_2 nanomaterials loaded on asphalt mixtures, the paper investigated that the effect of different adding processes of nano-TiO_2 on the photocatalytic degradation efficiency on asphalt mixture. First, the nano-TiO_2 was modified by surface modification, and its hydrophilicity was changed to lipophilicity. Then, the study discussed the effects of nano-TiO_2 adding processes and asphalt mixture types in the form of filler (dry processes) and in the form of asphalt additive (wet processes) on photocatalytic performance, through photocatalytic performance test. The test results show that: the degradation efficiency of porous asphalt mixture was 12% higher than that of dense gradation asphalt mixture, because contact area between photocatalyst and UV light in porous asphalt mixture was higher. The photocatalytic degradation efficiency of dry processes was about 33%, which was higher than about 22% of being added into the asphalt mixture using wet processes. When dry processes was employed, the photocatalytic degradation efficiency increased with the nano-TiO_2 increasing, and reached the maximum of 35%, and then decreased. Through analysis, dry processes had a unique advantage that nano-TiO_2 and mineral powder homogeneously mixed, hindered the agglomeration of Nano-particles, which makes it better in the dispersion, and thus exhibits better photocatalytic performance.

1 Introduction

Automobile exhaust gas is one of the main pollution sources in the atmospheric environment. Nitrogen oxides (NO_X) is one of the most harmful and the most difficult automobile exhaust gas to deal, so it is very important to purify and control NO_X in automobile exhaust gas. Nano-TiO_2 photocatalytic technology has become the most promising environmental pollution control technology with its non-toxic, low energy consumption, high degradation efficiency, mild reactive conditions and less secondary pollution. It has great application potentiality to remove the NO_X gas.

K. Zhang et al. (eds.), *Testing and Characterization of Asphalt Materials and Pavement Structures*, Sustainable Civil Infrastructures,
https://doi.org/10.1007/978-3-319-95789-0_8

At present, the nano-TiO_2 loaded onto the organic carrier, the group mainly studied the decontamination, deodorant, antibacterial and other aspects of the role, for example, load to the polyester, polyurethane fiber (Qian 2004; Li et al. 2004). As the Nano-materials has a strong redox capacity under the action of purple light, it can oxidize the NO of automobile exhaust to nitrates, so it play a role that make the air clean (Fujishima et al. 2000). Therefore, it is important to study that add the nano-TiO_2 to degrade the NO on the pavement. The effect of relative humidity on the NO degradation of Nano-coating was studied by Maggos et al. It was found that the degradation efficiency would decrease with increasing relative humidity in the range of 20–50% (Maggos et al. 2007). The effect of UV radiation intensity on NO degradation of nano-TiO_2 coating was studied by Chen et al. It was found that the degradation efficiency increased with the increase of the intensity of ultraviolet radiation (Chen and Chu 2011). The degradation efficiency of nano-TiO_2 at different heights from the road was studied by Ballari et al. It was found that the NO degradation efficiency at 5 cm from the road higher than 1/3 that at the 150 cm (Ballari and Brouwers 2013). The application of Nano-particle and rubber powder in improving the effect of NO degradation was studied by Liu et al. It was found that the above-mentioned composite technology could significantly improve its degradation efficiency (Liu et al. 2015). Nowadays, the research on the efficiency of nanotechnology depredating NO has been abundant, but there are still some shortcomings in the study of the addition process of photo catalyst. Therefore, this study will help to promote the application of nano-TiO_2 in practical engineering.

In this paper, the organic mixture of asphalt material is mainly used to discuss the photocatalytic degradation performance of the catalyst with the asphalt mixture as carrier. For the asphalt mixture, two kinds of blending mode are mainly discussed: one is the use of photocatalyst-containing asphalt mixing mixture (wet way); the other is the so-called dry addition, which is, the nano-TiO_2 powder mixed evenly with mineral filler, and then mixed with other components. The test also discusses the difference in photocatalytic performance between two asphalt mixtures (dense graded asphalt mixture and open graded asphalt mixture).

2 Fundamental Principle of TiO_2 Photocatalytic Degradation

TiO_2 is a new type of semiconductor (n-type) material with wide band gap, which Chemical properties is stable. Since the semiconductor energy band is discontinuous, under the light within a certain wavelength range, the material can absorb the light energy whose wavelength is above its bandgap width, and an electron transition occurs to form a hole (h^+)—electron (e^-) pair, then produce a strong free radicals and superoxide ions which is easy to organic matter and harmful gases catalytic decomposition (Zhang et al. 2012; Xu et al. 2011; Gao et al. 2013).

According to this theory, when added the nano-TiO_2 to road material and under the light conditions, TiO_2 can be changed to a photo catalyst. The vehicle emissions, like Carbon monoxide CO, hydrocarbon HC and nitrogen oxides NO_X, are catalyzed and oxidized. The forming of the corresponding carbonate and nitrate adsorption on the road surface, which these would be washed away by rain. The decomposition principle can be expressed as follows:

$$CO + O_2 \xrightarrow{\text{UV,} \quad \text{Nano-TiO}_2} CO_2$$

$$HC + O_2 \xrightarrow{\text{UV,} \quad \text{Nano-TiO}_2} H_2O + CO_3^{2-}$$

$$NO_X + O_2 \xrightarrow{\text{UV,} \quad \text{Nano-TiO}_2} NO_3^-$$

3 Test Materials and Equipment

3.1 Raw Materials

Base asphalt was Qingdao A-90 asphalt. The performance was shown in Table 1. Aggregate was coarse aggregate selected neutral basalt, according to 13.2, 9.5 and 4.75 mm to separate storage after screening. The performance was shown in Table 2. There were two kinds of fine aggregate dolomite chips and river sand. The filler was dolomite powder. The proportions of stone chips, river sand and mineral powder were shown in Table 3.

Photocatalytic materials was the technical indicators of the anatase nano-TiO_2, which absorb relatively strong UV used in this experiment shown in Table 4.

Table 1. Basic performance of asphalt

Asphalt type	Penetration/0.1 mm (25 °C)	Ductility/cm		Elastic recovery (25 °C)/%	Softening point/°C	Penetration index (PI)
		5 °C	25 °C			
A-90	85.2	5.7	>120	12	48.6	−1.57

Table 2. Basic technical performance of coarse aggregate

Stone type	Sturdiness/%	Los Angeles abrasion/%	Crushing value/%	Impact value/%	Water absorption/%	Elongated particle contents/%
Technical requirements	≤12	≤28	≤26	≤28	≤2.0	≤15
Basalt	0.82	9.80	10.61	9.0	0.78	4.7

Table 3. Screening results of fine aggregate and filler

Sieve size (mm)	13.2	9.5	4.75	2.36	1.18	0.6	0.3	0.15	0.075
Stone chip pass rate (%)	100	100	98.3	54.9	3.92	25.8	16	12.6	8.4
River sand pass rate (%)	100	100	95.3	88.0	79.5	63.2	25.5	5.9	1.12
Mineral powder pass rate (%)	100	100	100	100	100	100	100	99.1	68

Table 4. Technical indicators of anatase TiO_2

Technical indicators	Test results	Specification value
Specific surface area/(m^2/g)	68	50–90
pH	5.3	2–6
Particle size/nm	20	15–30
Loss on drying (%)	1.1	$\leq$1.5
Loss on ignition (%)	0.5	$\leq$2.0
Content (%)	99.8	>99.8

3.2 Test Equipment

The photocatalytic reactor of self—made flow tube type was used in the experiment. The material of the reactor was a heat-resistant glass with better UV transmittance. The reactor was a cylindrical with a length of 500 mm, an inner diameter of 60 mm, and the total volume of 1413 mL (Fig. 1). The test was divided into two parts: laboratory simulation air source test and actual gas test. The process was shown in Figs. 2 and 3. The test selected the black light lamp as the UV light source of photocatalytic reaction. According to spectrophotometric method of naphthalene ethylenediamine hydrochloride (Take the NO_2 m) of GB 17096-1997, the nitrogen oxides were measured at a wavelength of 540 nm using 752-type UV grating spectrophotometer.

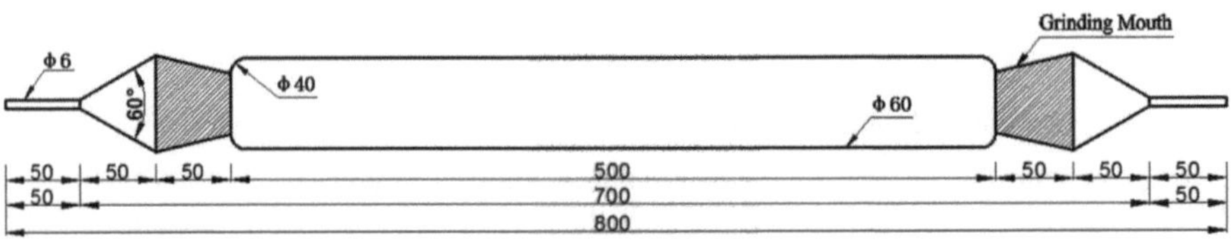

Fig. 1. Photocatalytic reactor dimensions unit: mm

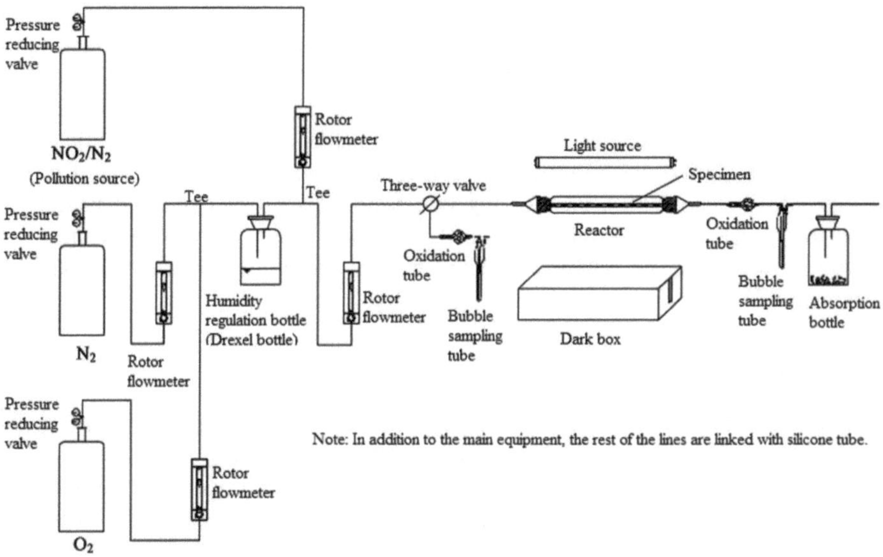

Fig. 2. Simulated air supply test flow chart

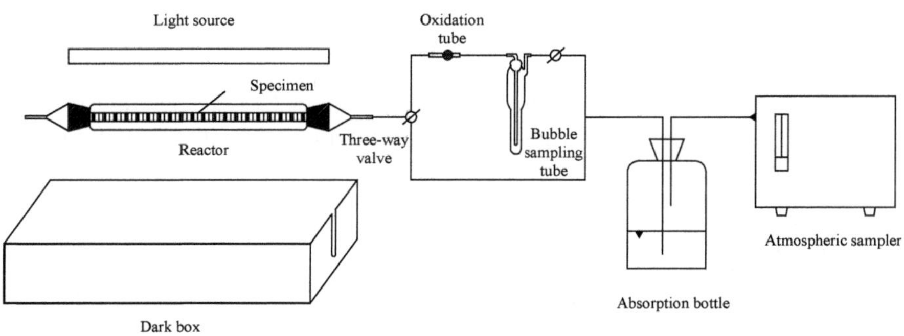

Fig. 3. Actual air supply test flow chart

4 Laboratory Test

4.1 Modification of Nano-TiO_2

The asphalt with lipophilic and hydrophobic is an organic macromolecule polymer which is composed a variety of mixtures, while the nano-TiO_2 with hydrophilicity and oleophobic properties is a small molecule inorganic material. So the compatibility between the two is poor. In order to improve the compatibility between the two, the nano-TiO_2 must be surface treatment from the hydrophilic into lipophilic, so that the nano-TiO_2 can be evenly and stably dispersed in the asphalt matrix, when under the premise of imparting photocatalytic performance does not change the physical and mechanical properties of the asphalt itself.

The trial uses octadecyltrimethylammonium bromide (OTAB) as modifier. It is a long-chain cationic surfactant, since the polar end has a positive charge, and the carbon chain of the non-polar end is longer, it can be adsorbed on the surface of TiO_2 under the action of electrostatic attraction. The longer carbons chain play a steric hindrance effect and make it uniformly dispersed in the asphalt matrix. The surface modification of nano-TiO_2 was carried out by ultrasonic dispersion. According to the reference (Ding and Fan 2001; Chen 2002), considering two factors like dispersion and parcels, the amount of OTAB was determined to be 0.5% of the mass of TiO_2. And the photocatalytic performance of nanometer TiO_2 before and after modification was also determined.

4.2 Adding Ways of Nano-TiO_2

In order to simulate the compaction process of the roller after paving the asphalt mixture, this test would make the asphalt mixture into rutting board specimens in accordance with regulatory requirements. Taking into account operability of the actual construction, this test would put Nano-TiO_2 into the asphalt mixture using two methods.

a. Wet processes: First, the modified nano-TiO_2 witch accounting for 6% of the quality of asphalt will be added into hot-mix of petroleum asphalt to prepare modified asphalt. Second, the modified asphalt is used to mix the dense gradation asphalt mixture (AC-13I) and Open Graded Friction Course (OGFC), and testing its photocatalytic performance.
b. Dry processes: It was used to filler that modified nano-TiO_2 accounting for 3, 4, 6, 12.5, 20% of the amount of asphalt, and then mixing well with mineral powder to batching the mixture.

4.3 Experimental Program

Two types of mixes, AC-13I and OGFC were used, and the mixture grades were shown in Tables 5 and 6.

In accordance with regulatory requirements, this test would make asphalt mixture into rutting board specimens with the standard size of 300 mm × 300 mm × 55 mm. Then, test cutted it into cuboid with size of 30 mm × 35 mm × 300 mm, and testing photocatalytic degradation efficiency in a photocatalytic reactor. Photocatalytic test

Table 5. AC-13I asphalt mixture gradation

AC-13I	Different sieve size (mm) passing rate (%)										OAC
	16	13.2	9.5	4.75	2.36	1.18	0.6	0.3	0.15	0.075	(%)
Upper limit specification	100	100	85	68	50	38	28	20	15	8	6.5
Lower limit specification	100	90	68	38	24	15	10	7	5	4	4.5
Synthetic gradation	100	94.3	77.6	54.5	35.2	25.9	19.2	12.8	9.7	7.2	5.1

Table 6. OGFC asphalt mixture gradation

AC-13I	Different sieve size (mm) passing rate (%)										OAC (%)
	16	13.2	9.5	4.75	2.36	1.18	0.6	0.3	0.15	0.075	
Upper limit specification	100	100	80	30	22	18	15	12	8	6	5
Lower limit specification	100	90	60	12	10	6	4	3	3	2	4
Synthetic gradation	100	94.3	73.2	23	15.6	11.6	9.3	8.1	4.3	3.1	4.6

conditions: The light source is a black light. The light intensity is 440 μW/cm^2. The relative humidity is 63%. Concentration of NO_2 is 0.4 mg/m^3. The photocatalytic performance of the asphalt mixture was measured using a NO_2 of artificial formulated certain concentration.

5 Test Results and Analysis

5.1 Photocatalytic Performance of TiO_2 Before and After Modification

The purpose of surface modification of nano-TiO_2 was to make it uniform and stable dispersion in the asphalt-based material, so as to play the superior photocatalytic performance. The test results show that nano-TiO_2 particles have good lipophilicity after modification with cationic surfactant OTAB, so that it can be uniformly and stably dispersed in the asphalt substrate. However, it was necessary to discuss the photocatalytic performance of TiO_2 after modification, since the surface of photo catalyst particles was covered by surfactant, and it may cause loss of its photocatalytic performance. It was shown in Fig. 4 that the photocatalytic performance of nano-TiO_2 before and after OTAB modification. It can be seen from the figure, nano-TiO_2 photocatalytic performance did not change significantly after the surface modification.

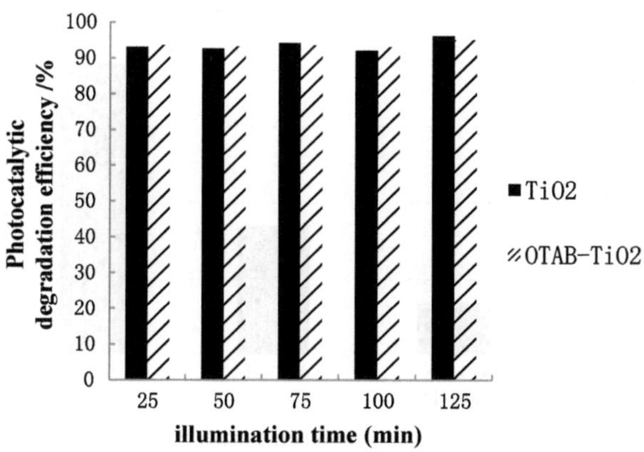

Fig. 4. Photocatalytic performance of nano-TiO_2 before and after surface modification

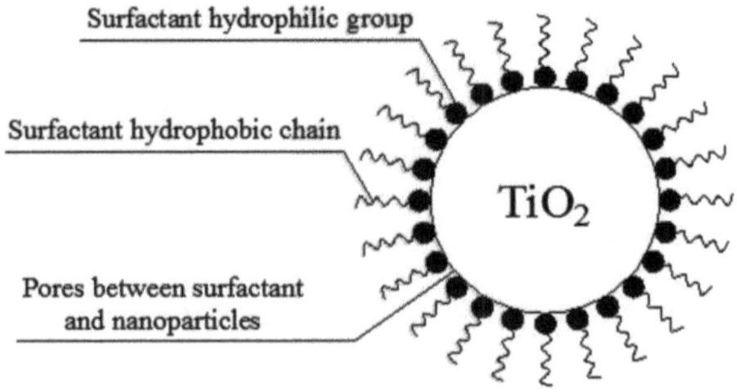

Fig. 5. Nano-TiO_2 particles after modification with cationic surfactant OTAB

Through analysis, it concluded that the surfactant was fixing the relative position of the nanoparticles to prevent agglomeration of nanoparticles, due to the surfactant fails to completely cover the surface of the nanoparticles to form a complete core-shell structure. It were shown in Fig. 5.

5.2 Photocatalytic Performance of Different Blending Methods

The photocatalytic properties of two types of asphalt mixture were discussed using the wet method. The test results were shown in Fig. 6. It can be seen from the figure that both of the conditions show a lower photocatalytic degradation performance. Wherein, the performance of the porous asphalt mixture was higher than that of the graded asphalt mixture. It was concluded that the porous asphalt mixture has more pores, and the contact area between the photo catalyst and the ultraviolet light is larger than that of the dense-graded asphalt mixture. So, the former showed a higher photocatalytic performance. However, the asphalt belongs to the cementation material which photocatalytic performance is not high, so that has only a weak photocatalytic degradation performance for asphalt mixture.

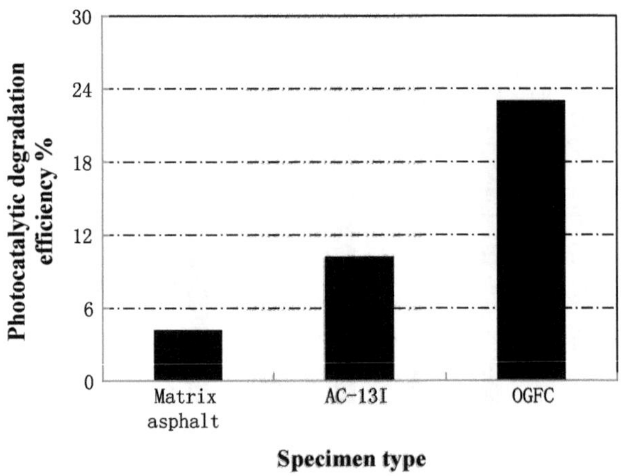

Fig. 6. Photocatalytic performance of asphalt mixture (wet method)

It is shown in Fig. 7 that photocatalytic degradation curve of preparing porous asphalt mixture using dry progress. It can be seen from the figure that the photocatalytic degradation efficiency of the mixture increases with the increase of nano-TiO_2. However, the photocatalytic degradation efficiency increases with the addition of nano-TiO_2 after adding a certain amount to a certain value. And then continuing increase amount of nano-TiO_2, the photocatalytic degradation efficiency have a trend of decline. As the Nano-powder has a huge specific surface area, there is a strong nanoaction between the nanoparticles to reunite easily. When the amount of nano-TiO_2 is adding too high, the mixing is not uniform with the mineral powder. This leads to that agglomeration of nanoparticles can't be spreaded out evenly during mix asphalt mixture, so that the photocatalytic degradation efficiency is reduced.

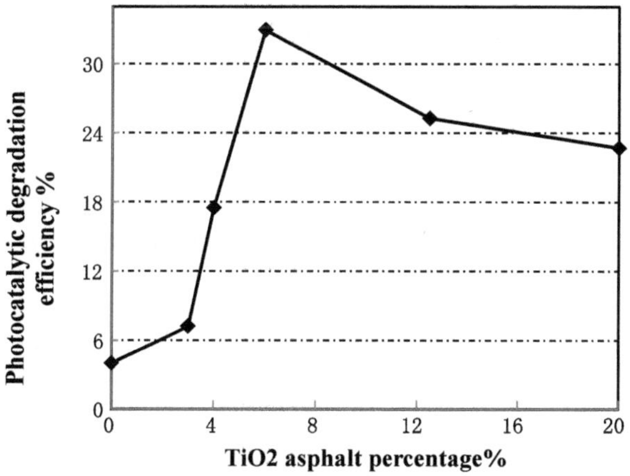

Fig. 7. Photocatalytic performance of asphalt mixture (dry method)

It was found that the photocatalytic degradation efficiency of the dry method prepared mixture was slightly higher than that of the former, through compare photocatalytic properties of porous asphalt mixture using two methods to be prepared. When dry method was used, dispersion of nanoparticles were more uniform than that using wet method to be prepared in mixture. Thus, this method exhibits a higher photocatalytic.

The test also tested the physical properties of asphalt mixture adopting two method to be prepared. The test results were shown in Table 7. It can be seen that nano-TiO_2 has a certain influence for the physical properties of asphalt mixture under having the premise of photocatalytic performance.

Table 7. Physical properties of asphalt mixture

Sample	Stability (KN)	Flow value (mm)
Original mix	2.56	34.3
6% (dry method)	2.28	29.9
6% (wet method)	2.23	25.4

6 Conclusion

Using OTAB surface modification method, the lipophilicity treatment of nano-TiO_2 can better play the photocatalytic performance without changing the original physical properties of original asphalt. As the nano-TiO_2 particles with negative charge, OTAB with a positive charge of the polar groups adsorbed on the surface of the particles under electrostatic interaction, while the non-polar long chain of carbon into the organic solution, so as to enhance the consistency between asphalt and nano-TiO_2.

The content of nano-TiO_2 has a certain influence on the photocatalytic degradation efficiency, and the photocatalytic efficiency increases with the increase of the content in a certain range. When the content is 6%, the photocatalytic degradation efficiency is the highest 25%, and the photocatalytic degradation efficiency decreases with increasing of the dosage, and the physical properties of the asphalt are deteriorated.

The type of asphalt mixture is one of the factors affecting the photocatalytic degradation efficiency. Since the photocatalytic performance is mainly related to the contact area between the actinic agent and the ultraviolet light, the porous asphalt mixture exhibits better catalytic efficiency.

It was found that the photocatalytic degradation efficiency of the dry way prepared mixture was slightly better than that at wet one, and porous asphalt mixture had obviously better photocatalytic performance than traditional dense gradation asphalt mixture.

Acknowledgements. This work was supported by funding from Tianjin major science and technology research program (16ZXCXSF00110), Tianjin application and advance technologies program (15JCYBJ23100), Hebei province natural science fund projects (E2016202279), Tianjin high education development funds (20140918).

References

Ballari, M.M., Brouwers, H.J.H.: Full scale demonstration of air-purifying pavement. J. Hazard. Mater. Elsevier (2013). https://doi.org/10.1016/j.jhazmat.2013.02.012

Chen, M., Chu, J.-W.: NO_X photocatalytic degradation on active concrete road surface—from experiment to real-scale application. J. Cleaner Prod, Elsevier (2011). https://doi.org/10.1016/j.jclepro.2011.03.001

Chen, Z.: Technical Questions and Answers on Production of Titanium Dioxide. Chemical Industry Press (2002). ISBN: 9787502520977

Ding, Y., Fan, C.: Study on surface coating of nanometer titanium dioxide. Mod. Chem. Ind. CNKI (2001). https://doi.org/10.16606/j.cnki.issn0253-4320.2001.07.005

Fujishima, A., Rao, T.N., Tryk, D.A.: Titanium dioxide photocatalysis. J. Photochem. Photobiol. C: Photochem. Rev. Elsevier (2000). https://doi.org/10.1016/s1389-5567(00)00002-2

Gao, L., Yu, X., Liu, Q., Wang, H.: Research Progress of Photocatalytic Oxidation of Modified TiO_2 in Water Treatment. Environmental Science and Technology, CNKI (2013). https://doi.org/10.3969/j.issn.1674-4829.2013.05.016

Li, X., Wei, J., Li, Y.: Study on nano-antimicrobial composites of titanium dioxide/apatite. J. Funct. Mater. CNKI (2004).

Liu, W., Wang, S., Zhang, J., Fan, J.: (2015). Photocatalytic degradation of vehicle exhausts on asphalt pavement by TiO_2/rubber composite structure. Constr. Build. Mater. Elsevier. https://doi.org/10.1016/j.conbuildmat.2015.02.034

Maggos, T., Bartzis, J.G., Leva, P., Kotzias, D.: Application of photocatalytic technology for NO_X removal. Appl. Phys. A J. Springer (2007). https://doi.org/10.1007/s00339-007-4033-6

Qian, B.: Application of nanomaterials in petrochemical industry. New Chem. Mater. CNKI (2004). https://doi.org/10.3969/j.issn.1006-3536.2004.04.007

Xu, H., Liu, L., Sun, L., Li, J.: Application of nanometer titanium dioxide in actual road engineering. Highw. Eng. CNKI (2011).

Zhang, W., Zou, Y., Sun, G., Wang, F.: Study on influencing factors of photocatalytic performance of titanium dioxide asphalt mixture. J. Wuhan Univ. Technol. CNKI (2012). https://doi.org/10.3963/j.issn.1671-4431.2012.03.009

Evaluating the Effect of Solar Radiations on the Performance of Asphalt Concrete Pavements

Amal Abdelaziz(✉) and Chun-Hsing Ho

Department of Civil Engineering, Construction Management & Environmental Engineering, Northern Arizona University, Flagstaff, AZ, USA
{aa2399, chun-hsing.ho}@nau.edu

Abstract. In recent years, severe weather events have impacted the performance of transportation infrastructures. The increase in temperature and heatwaves, lead to the early deterioration of pavement structures. The purpose of this paper is to study the effect of solar radiations on the behavior of flexible pavements. A project was implemented in Flagstaff, Arizona, to evaluate the impact of solar radiations on asphalt concrete pavements. The analysis was performed using similar road sections with shaded and unshaded areas. A three-dimensional model of the study area was built using MicroStation, to generate solar analysis study using specified time periods. To validate the numerical analysis, field visits were conducted to measure pavement surface temperatures, and evaluate distress types along pavement surfaces for areas with solar exposure and areas without sun exposure. Bending Beam Rheometer (BBR) tests were performed to determine the stiffness of the asphalt mixtures. ANSYS, a finite element analysis software, was used to analyze the critical tensile and compressive strains exerted on the pavement. According to the results, pavement sections exposed to solar radiations, experience a significantly higher level of critical tensile and compressive strains compared to shaded pavement areas. Based on test findings and analysis results, the paper concluded that solar radiations lead to the early deterioration of asphalt pavements by increasing fatigue distress and rutting.

Keywords: Pavement · Asphalt · Solar radiation · Critical strains
Fatigue · Rutting

1 Introduction

Urban Heat Islands (UHI) describes areas that are significantly warmer than nearby rural areas. Roads and pavements cover a significant percentage of urban surfaces; as a result, paved roads are considered one of the main contributors to urban heat islands (Yang et al. 2016). Pavement surfaces cause urban heat islands due to their low solar reflectance and high heat storage capacity (Chen et al. 2017). The high thermal capacity of pavements, allow them to absorb huge amount of solar energy, causing high temperatures along the surface (Wan et al. 2012). High temperatures could significantly shorten service life of pavement by increasing the rate of distress. As the temperature

K. Zhang et al. (eds.), *Testing and Characterization of Asphalt Materials and Pavement Structures*, Sustainable Civil Infrastructures,
https://doi.org/10.1007/978-3-319-95789-0_9

rises, asphalt binders are degraded by volatilization and oxidation, resulting in a rapid hardening of the pavement, which leads to the formation of cracks and severe damage to the surface (McPherson and Muchnick 2005).

Previous research showed that Ultraviolet radiations affect the upper layer of asphalt pavements and progresses the aging of the pavement, by influencing the low temperature ductility and resistance of cracks characteristics of asphalt binders and mixtures. Thus, it is essential to consider not only the effect of thermal stress on asphalt mixtures due to solar radiations but also the UV radiations effect as well. To mitigate the effect of solar radiations on asphalt pavements, researchers used carbon black and layered double hydroxide modifiers. These modifiers reduce the impacts of solar radiations by reflecting and absorbing the UV radiations, resulting in an improvement in the UV ageing resistance characteristics (Hu et al. 2017).

In recent times, modeling solar radiations is becoming more common, particularly due to the increase in unpredicted weather events. Modeling solar radiations can be performed using different techniques and computer programs, which includes geospatial data methods such as ArcGIS software, weather generators tools or global solar radiation tools that are based on deterministic or stochastic methods (Donatelli et al. 2006). In this study, solar radiations were estimated using MicroStation Software. MicroStation is a modeling, documentation and visualization software that can be used for two dimensional and three-dimensional designs. Solar radiation analysis in MicroStation can be performed over a user-defined period. Moreover, the software is capable of creating animations of solar studies from day to night, taking into account solar intensity and shadowing effects.

Fatigue cracking due to repeated traffic loads and rutting are two major modes of failure in asphalt pavements. Both types of pavement distresses occur mainly due to the applied traffic load and environmental conditions (Gogoi et al. 2013). The exposure of asphalt to solar radiations simulates the atmospheric degradation process of asphalt binders, resulting in the occurrence of fatigue cracking (Lins et al. 2008). Fatigue distress affects the strength and stiffness properties of asphalt mixtures (Jiangmiao and Guilian 2013). In addition, fatigue loading could lead to a reduction in the pavement service life, by allowing moisture to infiltrate through the cracks, resulting in large potholes (Gao et al. 2012). Rutting or permanent deformation can also influence the service life of asphalt pavement, especially during warm climates (Hossain and Zaman 2012). High temperatures increase fatigue cracking and rutting along the pavement surface, by influencing the elastic modulus of the top asphalt layer. Previous studies showed that temperature and elastic modulus of asphalt are inversely proportional. As the temperature increases, the elastic modulus of asphalt layer decreases (Behiry 2012). The reduction in the modulus or strength of asphalt increases the critical strain and shear stress exerted on the pavement, which makes the pavement surface more prone to deformation (Xiaodi et al. 2017).

The objective of this study is to investigate the influence of solar radiations and high temperatures on the performance of flexible pavements. The major concern of the study is to evaluate the effect of solar radiations on the intensity of fatigue cracking and rutting along asphalt concrete pavement surfaces. A three-dimensional model of the study area was created in MicroStation Software to simulate real world conditions in terms of shaded and unshaded areas. Bending Beam Rheometer tests were conducted to determine the creep compliance and stiffness values of the asphalt mixture used in the study. Finite element analysis was performed to determine the critical stresses exerted on the pavement surface at different temperatures. Fatigue and rutting lives were predicted based on the calculated critical stresses and strains.

2 Methodology

2.1 Study Area

In the study, six different road sections were used to evaluate the influence of solar radiations on asphalt concrete pavements. The study area is located at downtown Flagstaff, Arizona. The city of Flagstaff is located at an elevation of approximately 7000 ft, above the sea level. The climate is characterized by very dry, hot summers, and cold, snowy winters. The average temperature in Flagstaff is around 43.8 °F. The temperature in the summer gets as high as 80 °F. While, winter temperatures usually fall below 20 °F. Due to the high elevation of the city, temperature varies significantly during the day and night (Climate Flagstaff 2017). Thus, making it a suitable place to study the effect of temperature variations and solar radiations on pavement structures. The road sections were chosen to have the same geometric properties, layered pavement systems, and approximately equal average daily traffic. The shading of pavements in the study area is mainly due to the surrounding buildings.

2.2 Solar Radiation Modeling

A knowledge of the intensity of solar radiations received by pavement surfaces is essential to evaluate the influence of solar energy on pavement performance. To predict solar radiations absorbed by pavement, a three-dimensional model of the study area was built, based on the existing geometric features using MicroStation Software, as illustrated in Figs. 1 and 2. The advantages of using MicroStation for modeling purposes include its ability to generate solar analysis studies and shadowing effects based on the existing features of the study area. The software also provides animation tools to create images of shadow locations over specified period of time. Moreover, the solar animation in MicroStation includes the color and intensity of solar lightning.

A solar analysis study was generated for the model, using MicroStation. The study was done based on the exposure time of the pavement to solar radiations for a period of 24 h. A solar intensity of 1000 watts per square meters was used for the analysis. The results of the solar analysis study are presented in Figs. 3 and 4. Figure 3 represents the amount of time, for which the model is exposed to solar radiations in a 1-day period. The difference in the amount of solar radiations received by pavement is due to the shadows produced by the surrounding buildings. Pavement sections covered by the shadow of buildings, received less solar radiations than sections exposed to sunlight. The amount of time in which pavement surfaces were exposed to solar radiations varied between 5 and 13 h daily. Figure 4 illustrates the amount of solar radiation received by pavement in kilowatt-hours per square meters. Pavement surfaces received solar radiations between 2.9 and 7.4 KWh/m^2.

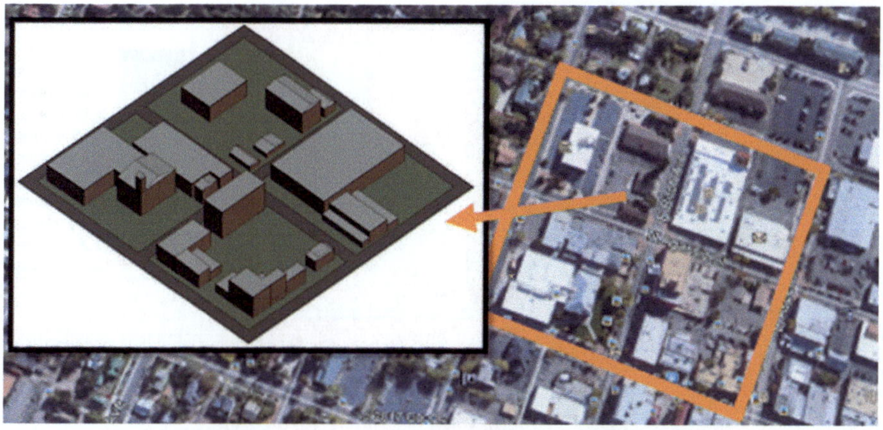

Fig. 1. 3D view of the MicroStation Software Model

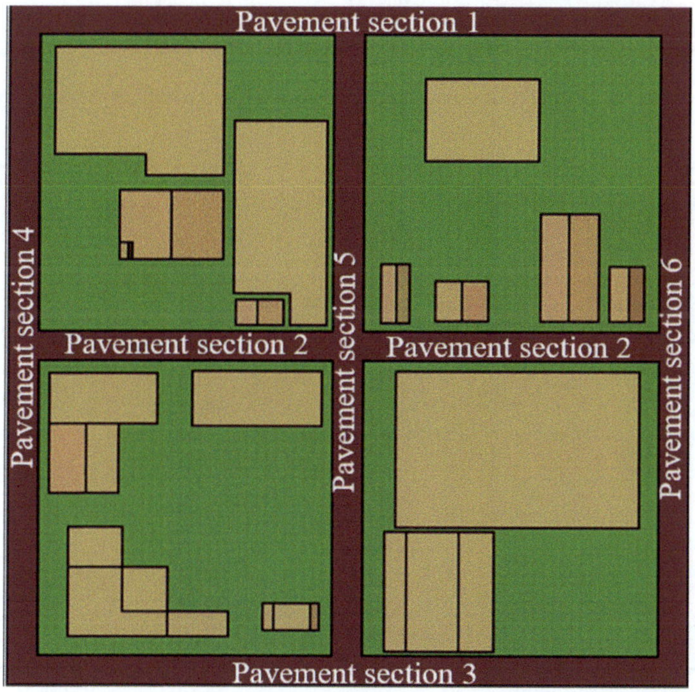

Fig. 2. Top view of the MicroStation Software Model

2.3 Field Data

Field visits were conducted to validate the software simulation results and to collect temperature data. It was observed that pavement surfaces exposed to sunlight have relatively more distress than shadowed pavement sections. Two main types of

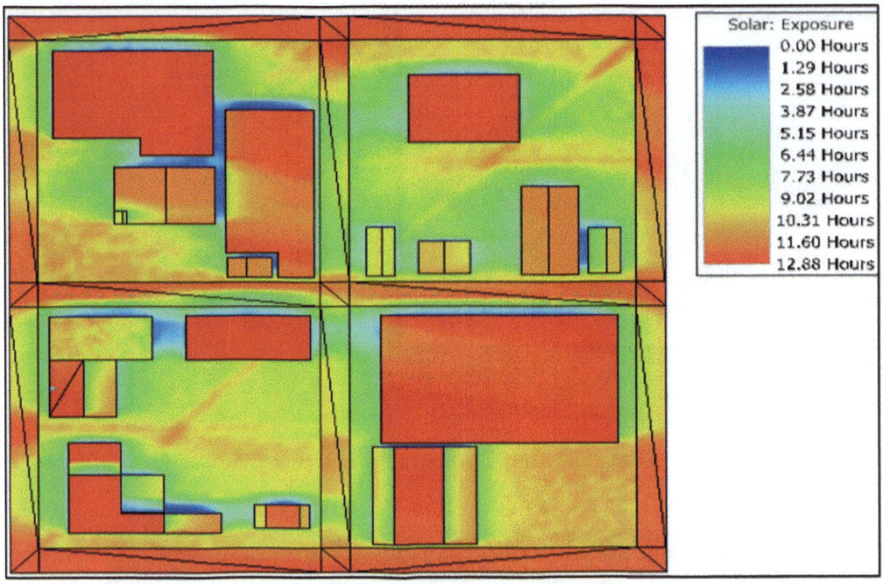

Fig. 3. Amount of solar exposure received by pavement surface

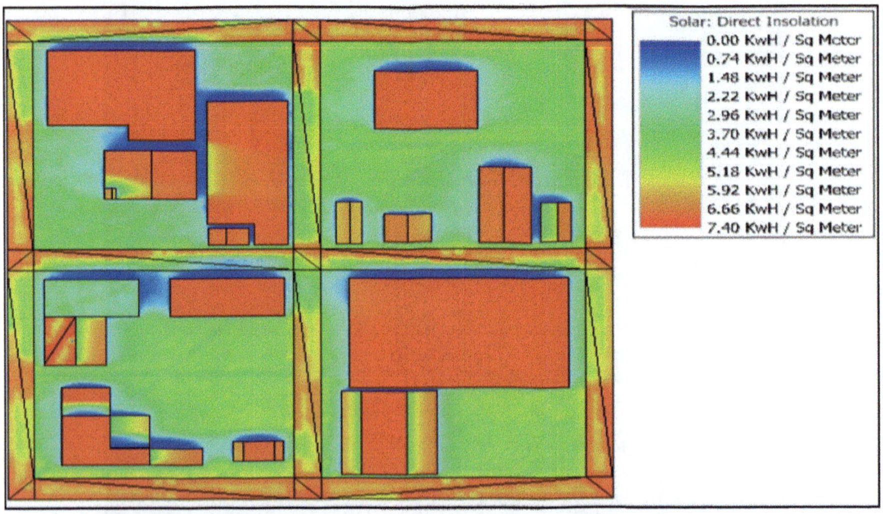

Fig. 4. Amount of solar radiation received by pavement surface

pavement distress were found: fatigue and thermal cracking. Pavement surface temperatures were measured using a thermal gun for the same day chosen to simulate solar radiation analysis in MicroStation. The highest pavement surface temperature recorded was 140 °F, and it was measured at noon. The average measured temperature of shaded pavement sections is equal to 80 °F. While, the average temperature of pavement surfaces exposed to sun is equal to 130 °F. The temperatures at the depth of asphalt pavement were determined based on Eq. 1 (Ho and Romero 2009).

$$T_{Pav} \times 0.859 + (0.002 - 0.0007 \times T_{air}) \times D + 0.17 \quad (1)$$

where:
T_{pav} = temperature of pavement at calculated depth (°C),
T_{air} = low air temperature (°C), and
D = depth (mm).

2.4 Bending Beam Rheometer Testing

Bending Beam Rheometer (BBR) tests were performed at temperatures of 12 °C, −18 °C, and −24 °C to determine creep compliances and stiffness values of the asphalt mixture used in the pavement test sections. The procedures used in BBR tests are based on the American Association of State Highway and Transportation Officials standard T313 for Determining the Flexural Creep Stiffness of Asphalt Binder Using the Bending Beam Rheometer (AASHTO 2009). Modifications were performed to test the creep compliances and stiffness values for the thin asphalt beams (Ho and Romero 2011, 2012). A BBR test is performed by applying a constant load on an asphalt beam, and measuring its corresponding deflection as a function of time. The BBR test measures the deflection of the beam at a time increment of 0.5 s. The stiffness values of the asphalt mixture at 60 s were used to determine the critical strains exerted on pavement layers.

2.5 Prediction of Stiffness Properties of Asphalt Mixtures at Different Temperatures

Stiffness of asphalt concrete mixtures is influenced by pavement surface temperature. Thus, the stiffness of asphalt pavement under shaded and unshaded areas are different. The LTPP Guide to Asphalt Temperature Prediction and Correction was used to estimate the value of elastic modulus at different pavement surface temperatures (Predictions 2000). The modulus of the asphalt layer was adjusted for different temperatures using the following model:

$$\text{ATAF} = 10^{\text{slope}\,(\text{Tr}-\text{Tm})} \quad (2)$$

where:
ATAF: calculation adjustment factor.
Slope: depends on the characteristics of the mix design including binder and aggregate properties. A default value of −0.021 was used to estimate the stiffness.
T_r: reference temperature, in Celsius.
T_m: measured temperature, in Celsius.

The estimated elastic modulus values for the asphalt layer were used along with the pavement structure properties in the finite element model to predict the tensile and compressive strains acting on the pavement layers. Tensile strain (ε_t) occurs at the bottom of the asphalt layer, and compressive strain (ε_c) occurs at the top of subgrade layer. Excessive tensile strain (ε_t) could result in pavement failure by increasing fatigue cracking. While, excessive compressive strain (ε_c) causes permanent deformation of the pavement surface due to rutting (Behiry, 2012).

2.6 Finite Element Analysis

A three-dimensional model was developed using ANSYS, a finite element software. Finite element analysis was performed to predict critical tensile strain at the bottom of the asphalt layer and compressive strain at the top of the subgrade layer, for different temperatures. To model the pavement structure, geometric properties, materials, mesh, and boundary conditions were assigned. The pavement structure consists of three layers: top asphalt concrete layer, gravel base layer, and subgrade layer with thicknesses of 14, 16, and 40 cm, respectively, as illustrated in Fig. 5. The structural layer properties of the model are presented in Table 1. The model was meshed using 6176 nodes and 900 elements. Elastic materials were used in the finite element analysis model. The boundary conditions were assigned by using horizontal and vertical constraints at the subgrade soil. The top two layers were modeled by restricting the displacement in all directions, except the vertical component. The responses of the pavement were evaluated by applying a tire pressure of 80 psi at the top layer.

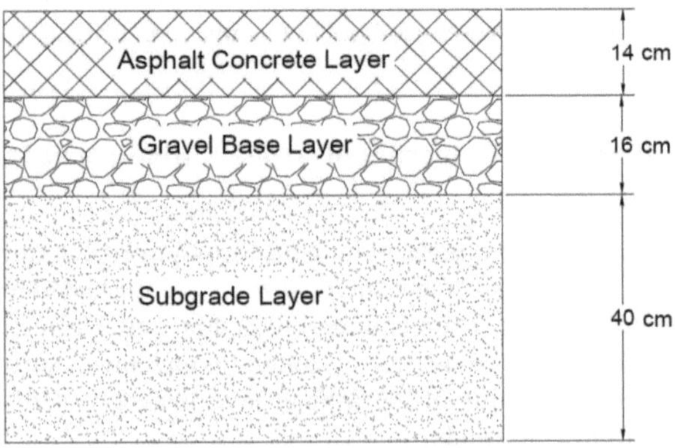

Fig. 5. The pavement structure

Table 1. Parameters of the layers of pavement structure

Layer name	Pavement thickness (cm)	Young's modulus/stiffness (MPa)	Poisson's ratio	Bulk modulus (MPa)	Shear modulus (MPa)
AC layer	14	526–2635	0.35	584–2928	195–976
Base layer	16	345	0.25	230	138
Subgrade	40	34.5	0.38	48	12.5

3 Pavement Damage Analysis: Fatigue Distress and Rutting

The following models were used to predict the number of load repetitions to prevent fatigue and rutting, based on the estimated elastic modulus values of asphalt and the critical strains acting on pavement.

Fatigue model
Fatigue damage was predicted using the following criteria:

$$N_f = f_1(\varepsilon_t)^{-f2}(E_1)^{-f3} \tag{3}$$

where:

N_f	is the allowable number of load repetitions to prevent fatigue cracking.
f_1, f_2, and f_3	are constants obtained from fatigue tests.
E_1	represents the stiffness of the asphalt overlay layer.

Rutting model
Damage due to rutting can be expressed as follows:

$$N_d = f_4(\varepsilon_c)^{-f5} \tag{4}$$

where:

N_d	is the allowable number of load repetitions to prevent rutting.
f_4 and f_5	are constants that are determined from road tests.

The values used for the road tests coefficients: f_1, f_2, f_3, f_4 and f_5 were determined based on the asphalt institute method specifications (Huang 2004), and are summarized in Table 2.

Table 2. Road test coefficients of fatigue life and rutting life models

Road test coefficient	f_1	f_2	f_3	f_4	f_5
value	0.0796	3.291	0.854	1.365×10^{-9}	0.0796

4 Results and Discussion

4.1 Influence of Solar Radiations on the Stiffness of Asphalt

To determine the influence of solar radiations on the stiffness or the strength of the asphalt mixture, the elastic modulus values were graphed against pavement surface temperatures. The relationship between solar radiations and stiffness of asphalt layer was found to be inversely proportional, as shown in Fig. 6. Pavement sections with higher solar exposure had a lower elastic modulus values compared to shaded sections. The elastic modulus of the asphalt layer at a pavement surface temperature of 140 °F was lower than the stiffness at 80 °F by 80%. The significant reduction in the asphalt stiffness due to the increase in temperature is critical as it increases the critical stresses applied on the pavement, resulting in a reduction in the overall lifespan of the pavement structure.

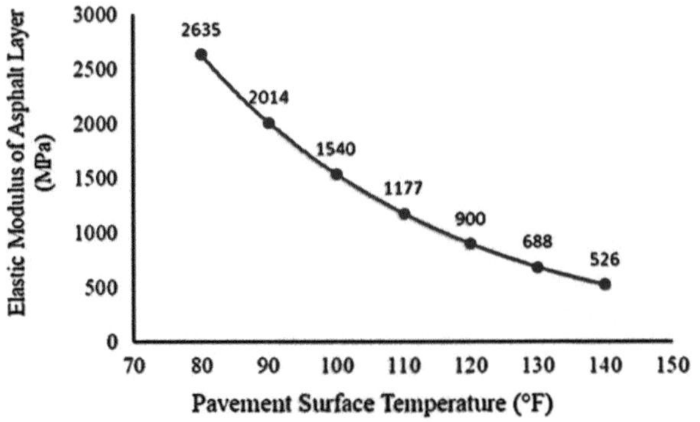

Fig. 6. Elastic modulus of asphalt layer versus temperature variations

4.2 Effect of Solar Radiations on Critical Strains

The results of ANSYS analysis of determining the tensile and compressive strains applied on the pavement structure are shown in Figs. 7 and 8. Based on the results, the maximum tensile and compressive strains occur at the bottom subgrade layer. To evaluate the influence of solar radiation on the critical strains exerted on asphalt pavement surface, critical strains were plotted against field-measured temperatures, as illustrated in Figs. 9 and 10. Tensile and compressive strains significantly increase with the increase in pavement temperature. The high correlation coefficient observed in the graphs indicates that the relationship between critical strain and the increase in temperature is significant and can be expressed using an exponential function. The results also show that the tensile strain at the bottom of the asphalt layer was higher than the compressive strain at the top of the subgrade layer.

4.3 Effect of Solar Radiations on Fatigue Distress and Rutting

To evaluate the effect of solar radiations on pavement distress, the number of allowed load repetitions to prevent fatigue cracking and rutting were graphed against different pavement surface temperatures, as shown in Figs. 11 and 12. The results show that fatigue life is shorter than rutting life. Thus, it can be concluded that fatigue cracking is the controlling parameter of the pavement structure failure. The relationship between fatigue life and the increase in temperature can be represented by a logarithmic function. While, the correlation between rutting life and temperature can be expressed by an exponential function. Based on the results, solar radiations could lead to a substantial reduction in rutting and fatigue lives. Pavement sections exposed to solar radiations are expected to have more than 50% shorter fatigue and rutting lives,

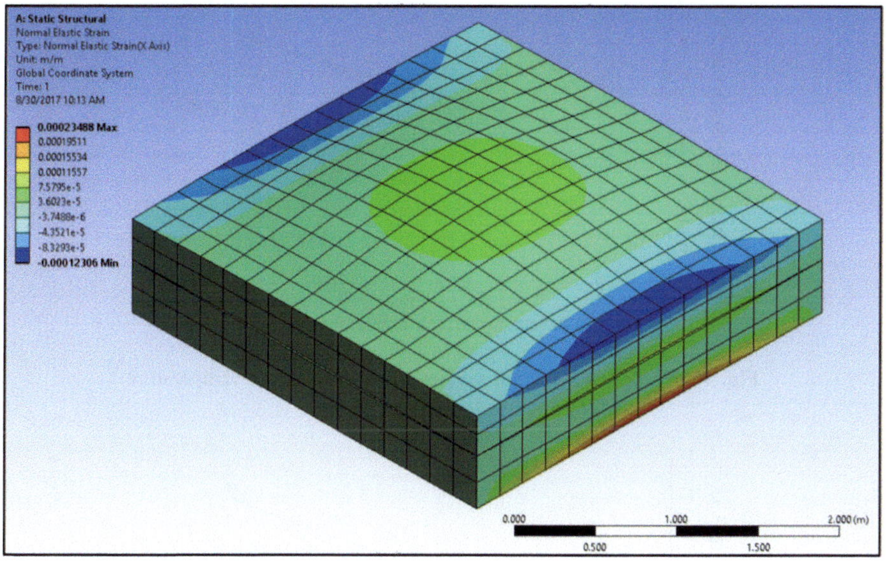

Fig. 7. Tensile strain applied on pavement structure

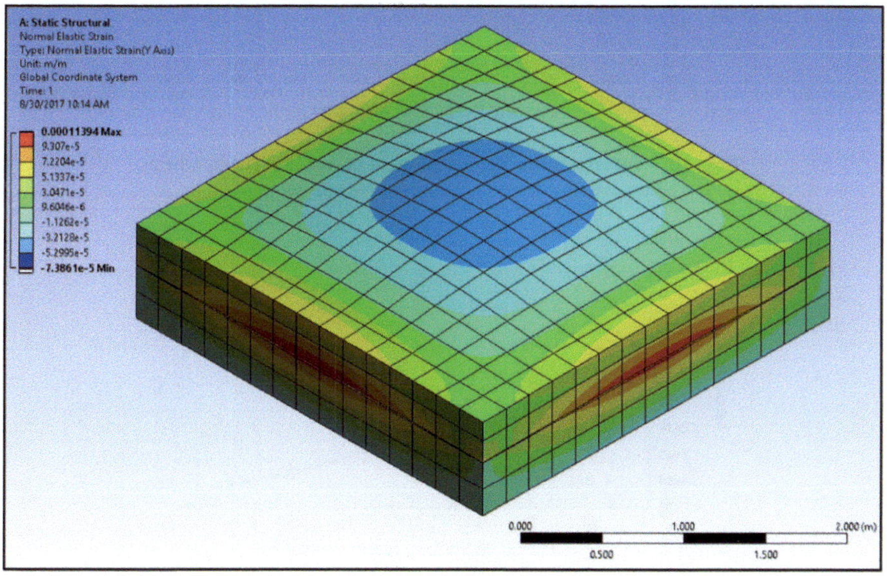

Fig. 8. Compressive strain applied on pavement structure

compared to areas with no solar exposure. Moreover, the fatigue life of pavement sections at shaded regions is predicted to be higher than areas with sun exposure by 66%.

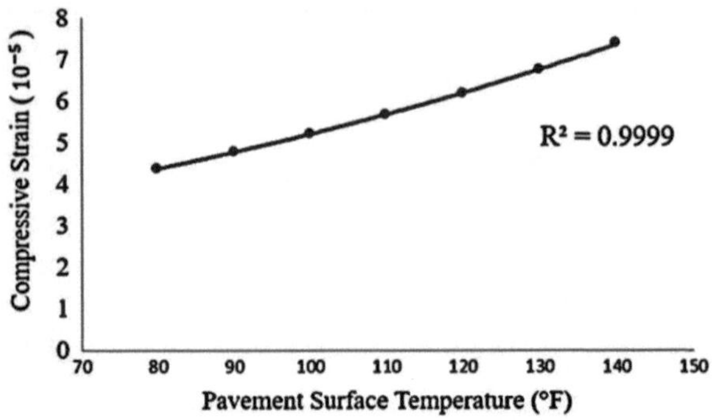

Fig. 9. Compressive strain versus pavement surface temperature

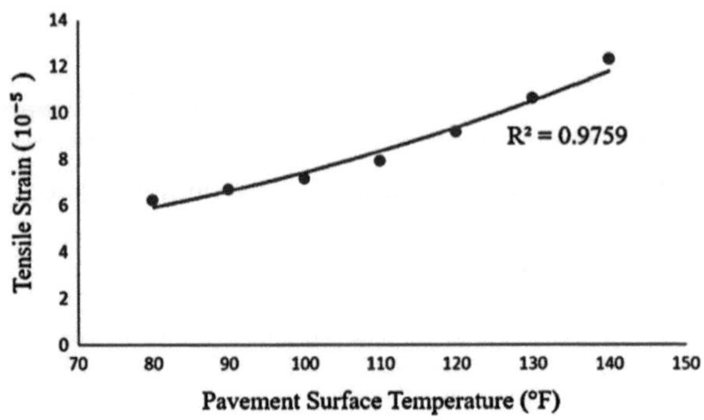

Fig. 10. Tensile strain versus pavement surface temperature

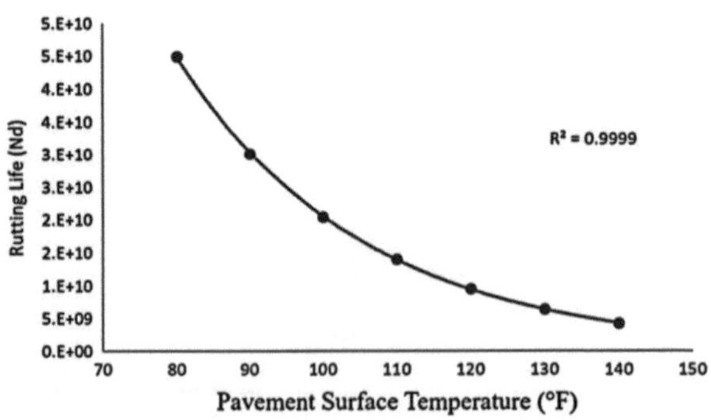

Fig. 11. Rutting Life versus temperature variations

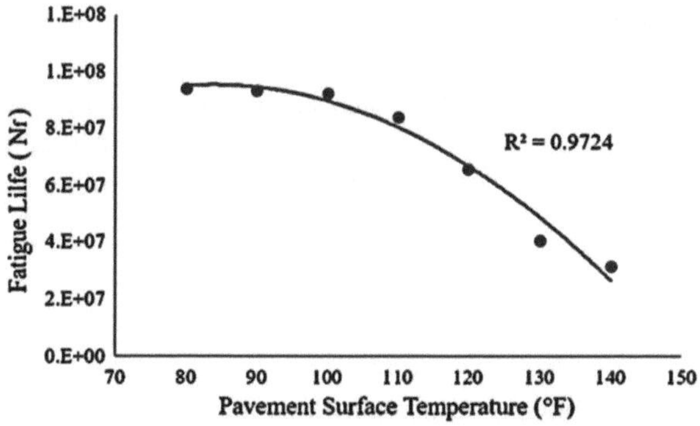

Fig. 12. Fatigue Life versus temperature variations

4.4 Statistical Analysis

The Influence of Solar Radiations on the Stiffness of Asphalt Layers

To determine the statistical significance of the relationship between solar radiations and the elastic modulus of the asphalt layer, a t-test was performed. The results of the t-test showed a t-statistic value of—4 and t-critical value of 2. Since, the magnitude of t-statistic is higher than t-critical, the null hypothesis can be rejected and it can be concluded that there is a significant relationship between solar radiations and stiffness of asphalt layers. In addition, this conclusion can be validated by observing the p-values. Both the one-tail and two-tails p-values are lower than 0.05, as a result it can be concluded that the effect of solar radiations on the elastic modulus of asphalt layer is significant.

The Relationship between Solar Radiations, Fatigue, and Rutting Lives

A one-way analysis of variances (ANOVA) approach was performed to determine the statistical significance of the relationship between solar radiations and the expected fatigue and rutting lives. The results of the statistical analysis are summarized in Tables 3 and 4. The statistical P-values calculated for the influence of pavement surface temperature on fatigue and rutting lives are 0.00192 and 0.00101, respectively. Based on the P-values obtained from the statistical analysis, it can be concluded that the increase in pavement surface temperature has a significant impact on fatigue and rutting lives.

Table 3. ANOVA analysis of the relationship between pavement surface temperature and fatigue life

Source of variation	Degrees of freedom	Sum squares	Mean square	F-value	P-value
Between groups	1	3.64×10^{15}	3.64×10^{15}	35.4	0.00192
Within groups	5	5.15×10^{14}	1.03×10^{14}		

Table 4. ANOVA analysis of the relationship between pavement surface temperature and rutting life

Source of variation	Degrees of freedom	Sum squares	Mean square	F-value	P-value
Between groups	1	1.16×10^{21}	1.16×10^{21}	46.8	0.00101
Within groups	5	1.24×10^{20}	2.48×10^{19}		

5 Conclusions

This study evaluated the effect of solar radiations on the performance of asphalt concrete pavements through laboratory tests and finite element analysis methods. Based on the results obtained from the study, the followings can be concluded:

- The elastic modulus of the top asphalt layer significantly decreases as pavement surface temperature increases.
- The relationship between tensile and compressive strains and the increase in pavement surface temperature can be represented by an exponential function.
- The maximum tensile and compressive strains occur at the bottom subgrade layer.
- The relationship between the increase in temperature and fatigue and rutting lives can be expressed using logarithmic and exponential functions, respectively.
- Solar radiations have a significant impact on the service life of asphalt pavements. Pavement sections exposed to solar radiations had a dramatically lower fatigue and rutting lives, compared to pavement sections at shaded regions.
- Considering the intensity of solar radiations received by pavement surfaces is essential to maximize the life span of pavement structures.

References

American Association of State Highway and Transportation Officials: Determining the Flexural Creep Stiffness of Asphalt Binder Using the Bending Beam Rheometer (BBR). Standard Specifications for Transportation Materials and Methods of Sampling and Testing T 313 (2009)

Behiry, A.E.A.E.M.: Fatigue and rutting lives in flexible pavement. Ain Shams Eng. J. **3**(4), 367–374 (2012)

Climate Flagstaff—Arizona: Retrieved 22 Aug 2017, from www.usclimatedata.com

Donatelli, M., Carlini, L., Bellocchi, G.: A software component for estimating solar radiation. Environ. Model Softw. **21**(3), 411–416 (2006)

Gao, L., Aguiar-Moya, J.P., Zhang, Z.: Bayesian analysis of heterogeneity in modeling of pavement fatigue cracking. J. Comput. Civil Eng. **26**(1), 37–43 (2012)

Gogoi, R., Das, A., Chakroborty, P.: Are fatigue and rutting distress modes related? Intern. J. Pavement Res. Technol. **6**(4), 269–273 (2013). https://doi.org/10.6135/ijprt.org.tw/2013.6(4).269

Ho, C.H., Romero, P.: Low design temperatures of asphalt pavements in dry-freeze regions: predicting by means of solar radiation, transient heat transfer, and finite element method. Transp. Res. Rec.: J. Transp. Res. Board **2127**, 60–71 (2009)

Ho, C.H., Romero, P.: Using asphalt mixture beams in the bending beam rheometer: experimental and numerical approach. J. Road Mater. Pavement Des. **12**(2), 293–314 (2011)

Ho, C.H., Romero, P.: Asphalt mixture beams used in the bending beam rheometer for quality control: Utah experience. J. Transp. Res. Board **2268**, 92–97 (2012)

Hossain, Z., Zaman, M.: Sensitivity of Oklahoma binders on dynamic modulus of asphalt mixes and distress functions. J. Mater. Civ. Eng. **24**(8), 1076–1088 (2012). https://doi.org/10.1061/(ASCE)MT.1943-5533.0000461

Huang, Y.H.: Pavement Analysis and Design. Pearson/Prentice Hall, Upper Saddle River, NJ (2004)

Jiangmiao, Y., Guilian, Z.: Asphalt Pavement fatigue cracking prediction model with mode factor. Intern. J. Pavement Res. Technol. **6**(2), 123–129 (2013). https://doi.org/10.6136/ijprt.org.tw/2013.6(2).123

Hu, J., Wu, S., Liu, Q., Hernández, M.I.G., Zeng, W., Xie, W.: Study of antiultraviolet asphalt modifiers and their antiageing effects. Adv. Mater. Sci. Eng. (2017). https://doi.org/10.1155/2017/9595239

Lins, V.F.C., Araújo, M.F.A.S., Yoshida, M.I., Ferraz, V.P., Andrada, D.M., Lameiras, F.S.: Photodegradation of hot-mix asphalt. Fuel **87**(15), 3254–3261 (2008)

McPherson, E.G., Muchnick, J.: Effects of street tree shade on asphalt concrete pavement performance (2005)

Predictions, F.T.: Adjustment Factors for Asphalt Pavement. Publication No. FHWA-RD-98-085, FHWA, Research and Development, McLean, VA, (2000)

Wan, W.C., Hien, W.N., Ping, T.P., Aloysius, A.Z.W.: A study on the effectiveness of heat mitigating pavement coatings in Singapore. J. Heat Island Inst. Intern. **7**, 2 (2012)

Xiaodi, H., Faruk, A.M., Jun, Z., Souliman, M.I., Walubita, L.F.: Effects of tire inclination (turning traffic) and dynamic loading on the pavement stress strain responses using 3-D finite element modeling. Intern. J. Pavement Res. Technol. **10**(4), 304–314 (2017). https://doi.org/10.1016/j.ijprt.2017.04.005

Yang, J., Wang, Z.H., Kaloush, K.E., Dylla, H.: Effect of pavement thermal properties on mitigating urban heat islands: a multi-scale modeling case study in Phoenix. Build. Environ. **108**, 110–121 (2016)

Effect of Overloaded Vehicles on Whole Life Cycle Cost of Flexible Pavements

Dawid Rys(✉) and Piotr Jaskula

Faculty of Civil and Environmental Engineering,
Gdansk University of Technology, Gdańsk, Poland
dawrys@pg.gda.pl

Abstract. The phenomenon of vehicle overloading—illegal exceeding of maximum legal weight of vehicles, is a serious problem both in developing and developed countries around the world. Overloaded vehicles occur less frequently in comparison to properly loaded vehicles but due to their greater potential to cause damage they significantly contribute to distress of pavement structure. As studies show, the number of overloaded vehicles increases when the control of traffic is insufficient. Weigh in Motion (WIM) systems significantly improve control level and contribute to decrease in the number of overloaded vehicles. Data delivered from WIM were used to perform statistical analysis of vehicle overloading in Poland. The average percentage of overloaded vehicles (OV) in Poland varies from 5% for roads with high enforcement level to 23% for roads where control is poor. Every weighed vehicle was considered in terms of exceeding maximum legal gross weight and maximum legal axle load. For each vehicle separately truck equivalency factors were calculated. Subsequently the relationship between average values of truck equivalency factors and percentages of overloaded vehicles was found. This relationship was used as a basis to determine the impact of overloaded vehicles on decrease in fatigue life of pavement structure (DFL) and increase factor IF, which expresses the extension of service period. It was proved that reduction of overloaded vehicles from 23 to 5% will contribute to increase in service period of pavement structures by factor 1.5. The life cycle cost analysis (LCCA) was performed for two levels of overloading OV = 23% and OV = 5%. The paper revealed that improvement of vehicle control and reduction of the percentage of overloaded vehicles from 23 to 5% will cause the reduction of whole life cost borne by road authority by 11%.

1 Introduction

1.1 Background

In Poland for the last two decades the traffic of heavy vehicles on the main roads network has been increasing rapidly. Especially in the vehicle class of five-axle trucks with semi-trailer it was noted that an average increase by five times occurred (GDDKiA 2015). After accession of Poland to the European Union in 2005 the maximum legal weight increased to 115 kN per single drive axle as a consequence of the European Union Council Directive 96/53/EC. It resulted in an increase in real axle loads of vehicles. Moreover,

K. Zhang et al. (eds.), *Testing and Characterization of Asphalt Materials and Pavement Structures*, Sustainable Civil Infrastructures,
https://doi.org/10.1007/978-3-319-95789-0_10

a significant fraction of vehicles exceed the legal limits of gross weight and axle loads (Szydlo and Wardega 2003; Zofka et al. 2014; Rys et al. 2016a, b).

The simplest statistical value used to describe the problem of vehicle overloading is the percentage of overloaded vehicles in the total number of trucks. It is observed that the percentage of overloaded vehicles is much higher in developing countries and can reach an extremely high level of 80% according to studies of Zhao et al. (2012) and Mulyono and Antameng (2010). In developed countries it is in lower range from 10 to 30% (Mohammadi and Shah 1992; Pais et al. 2013; Fiorillo and Ghosn 2014).

Overloaded vehicles have much greater potential to cause pavement distress in comparison to properly loaded vehicles (Jeongho Oh et al. 2007; Wang and Zhao 2016). The damaging effect of overloaded vehicles depends not only on their percentage in total number of trucks, but also on the probability distribution of vehicle loads (Mohammadi and Shah 1992; Rys et al. 2016a, b). The truck-induced damage cost varied significantly between thin and thick asphalt pavements (Wang and Zhao 2016). Wang and Zhao showed that a 1% increase in overweight trucks could cause a 1.8% reduction of pavement life. According to Al-Qadi et al. (2017) truck overloading causes greater pavement damage on secondary roads with low truck traffic volume.

The phenomena of vehicle overloading contribute to reduction in service period of pavements and bridges as well as an increase in the cost associated with maintaining, upgrading and replacing the highway infrastructure. According to Pais et al. (2013) the maintenance cost of road calculated per one vehicle is higher by 100% for overloaded vehicles compared to the cost of the same vehicle with legal loads. The infrastructure damage costs should be recovered by proper fee for overloaded vehicles (Dey et al. 2015; Al-Qadi et al. 2017). The estimated pavement damage cost varies significantly depending on truck traffic volume and highway type. However, the level of damage cost recovery fee depends on the effectiveness of detection of overloaded vehicles and it can be improved by regular vehicle control. Moreover, regular control reduces the percentage of overloaded vehicles. According to studies carried out by Tailor et al. (2000) in the state of New York (USA), the percentage of overloaded vehicles was reduced from 30 to 2% when enforcement level was increased.

The weigh in motion (WIM) system allows to improve vehicle control and contribute to decrease in the number of overloaded vehicles. However, in Europe WIM systems are used for vehicle preselection and those recognized as potentially overloaded have to be weighed again on legalized static scales. Thus the effectiveness of WIM systems depends significantly on proper localization of the WIM station in road network (Oskarbski and Kaszubowski 2016; Budzynski et al. 2017). The recent works (Burnos and Rys 2017; Burnos and Gajda 2016; Doupal and Calderara 2008) aim to improve WIM accuracy and solve related legislation problems in order to enable usage of WIM systems for automatic identification of overloaded vehicles and imposing of fines, which would significantly contribute to efficiency of control.

1.2 Objectives

The analyses presented in this paper use data from WIM in order to (1) determine the extent to which the reduction in number of overloaded vehicles contributes to extension of service period of flexible pavements and further (2) to develop a methodology of calculation of the profits from reduction of overloaded vehicles in whole Life Cycle Cost Analysis (LCCA).

2 Determination of the Impact of Overloaded Vehicles on Service Period of Flexible Pavements

2.1 Weigh in Motion Data Used in Analysis

The analysis were a part of wider research program concerning actualization of the polish catalogue of typical flexible and semi-ridig pavement structures (Judycki et al. 2017; Rys et al. 2016a, b; Pszczoła et al. 2016). Data from 11 WIM stations on Polish national roads and motorways were used to perform analysis. The WIM stations are equipped with bending plate sensors PAT DAW 100® (A2 and DK11) or with piezo-electric quartz sensors Kistler Lineas®. Automatic vehicle classification systems are also installed in all of the stations. The WIM stations can be classified as class B7 according to COST 323 WIM classification. The information about WIM data and measurement period are given in Table 1. The data considered in this study were collected from 2010 till 2016 but the period of measuring can differ for particular WIM stations (see Table 1). In all cases periods of data collection cover whole years. The WIM stations are installed for each traffic direction separately but for further analysis data from two directions from the same road section were analyzed together as for one measuring point.

Table 1. Time of WIM measurement and number of vehicles records used in analysis

WIM station	Years of measurements	Total number of records	
		Raw WIM records of all vehicles types	Records of trucks (after filtering process)
A2	2011–2012	8,411,233	2,580,957
S7	2012	3,661,002	417,185
DK4	2010–2015	23,519,182	3,299,388
DK7	2012–2015	18,264,592	1,927,799
DK11	2010–2011	2,968,179	606,668
DK22	2013	3,969,641	638,915
DK46	2011–2016	13,323,213	2,305,126
DK75	2012–2015	17,850,479	1,951,386
DK79	2013–2015	14,362,028	1,285,706
DK94	2013–2016	13,791,140	1,608,742
DK94c	2015–2016	7,778,335	1,271,672
Total		127,899,024	17,893,544

The raw WIM data were verified using a series of filters based on vehicle parameters (e.g. axle loads, total length, axle configurations etc.). The filters were set in accordance with the WIM Data Analyst's Manual (FHWA 2010) and NCHRP Report

538 (NCHRP 2005) as well as vehicle technical parameters review. The filtering process was focused on identifying and removing invalid records from the database. In total more than 127 million passes across all types of vehicles were recorded, including cars, vans etc., out of that almost 18 million records of trucks were used in the analysis.

2.2 Problem of Overloaded Vehicles in Poland

The basis for classification of a vehicle as overloaded was the European Union Council Directive 96/53/EC which specifies the legal limits of vehicles gross weights and axle loads. The value of legal limit of vehicle gross weight depends on the class of the vehicle, e.g. for 2-axles single truck unit it is equal to 18,000 kg and for 5-axles truck with semi-trailer it is equal to 40,000 kg. The legal limit of axle loads depends on the type of the axle (steering, drive etc.), the distance to neighboring axles (single, tandem, tridem) and the suspension type. For example, in EU for single drive axles the maximum axle load equals 115 kN and for other types of single axles it is equal to 100 kN.

In the analysis, each vehicle was checked and marked if overloaded with algorithms developed specifically for this study at the Gdansk University of Technology. A vehicle was treated as overloaded according to EU Council Directive 96/53/EC in the following cases:

- The vehicle gross weight was greater than the legal limit;
- The load of a single or tandem or tridem axle was greater than legal limit;
- Both the gross weight and the axle load were greater than legal limits.

The percentage of overloaded vehicles (abbreviated further as OV) was calculated for each station and each year of measurement. The annual average percentage of overloaded vehicles and the range of its variation is presented in Fig. 1.

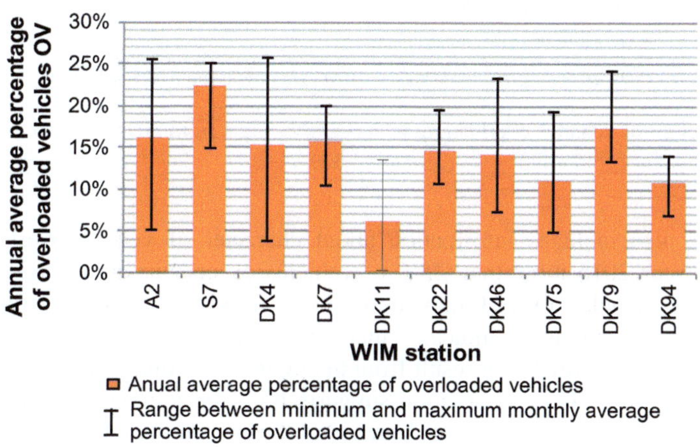

Fig. 1. Annual average percentage of overloaded vehicles OV in selected WIM stations in Poland

As it can be seen in Fig. 1 the annual average percentage of overloaded vehicles ranges from 3 to 23% depending on station and year. The maximum variations for one station do not exceed 7% over individual years. A more detailed analysis showed that the percentage of overloaded vehicles varies between days, weeks and months and was described in Rys et al. (2017). The level of enforcement can impact on this statistic. The interview with officials of the polish Road Transport Inspectorate was carried out and it provided the information, that. control on DK11 is carried out almost every day, while on S7 control is hardly ever performed. More detailed data about the frequency of controls was not available due to the confidentiality of such information. For stations where measurements were performed for several following years, like DK46, it was observed that the percentage of overloaded vehicles in first years after WIM system installation gradually decreased, which implies that installation of WIM station has a preventive effect. However, if control on legalized scales is too rare, the phenomenon of vehicle overloading intensifies, which is visible from data presented in Fig. 2 for an example station DK46. Similar changes in percentage of overloaded vehicles were observed in stations DK7, DK75 and DK94.

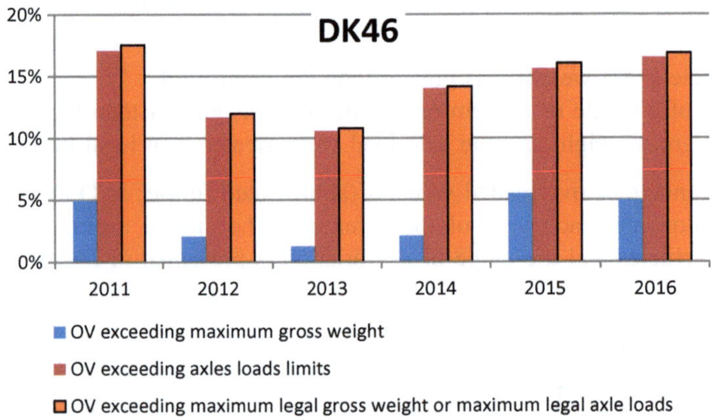

Fig. 2. Structure of vehicles overloading in successive years from 2011 to 2016 for an example station DK46

Figure 2 also includes information about character of vehicle overloading. For example in year 2015 more than 15% of vehicles exceed maximum axle load limits while 5% of vehicles exceed maximum gross weight. Similar disproportion was observed for all stations. It indicates that very often freight inside vehicles is not properly distributed, which can result from negligence and unawareness of drivers and transport companies. This observation suggests that—beside vehicle control—education of road users will contribute to decrease in the percentage of overloaded vehicles as well. To summarize the above discussion, it seems reasonable that increase in enforcement level and increase in awareness of road users can result in decrease in the percentage of overloaded vehicles to the level of 5%.

2.3 Effect of Overloaded Vehicles on Load Equivalency Factors and Truck Equivalency Factors

Truck Equivalency Factor (TF) characterizes the damaging effect of a given truck on pavement structures. The detailed procedure of calculating TF for this analysis is given in Rys et al. (2017). The fourth power equation was used to calculate load equivalency factors for particular axles in each vehicle separately and subsequently for every vehicle. As it was proved in work of Judycki (2010), fourth power equation provides similar load equivalency factors for flexible pavements as load equivalency factors determined on the basis of fatigue criteria. Damaging effects of single, dual and triple axles were considered separately according to Judycki (2006). Truck equivalency factors TF were calculated for every properly weighted vehicle and each of those vehicles was verified in terms of overloading. This approach allowed to calculate average values for different periods of time. The average truck equivalency factors TF and average percentage of overloaded vehicles OV were calculated for particular WIM stations, taking into account the whole period of analysis. The relationship between OV and TF is given in Fig. 3. More detailed analyses and relationships obtained for particular stations in monthly periods are given in Rys et al. (2017).

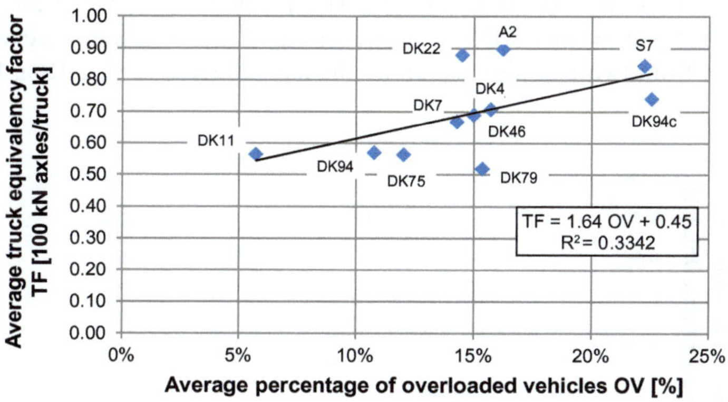

Fig. 3. Relationship between average truck equivalency factors TF and average percentage of overloaded vehicles OV obtained for particular WIM stations

It is clearly visible in Fig. 3 that an increase in percentage of overloaded vehicles OV causes an increase in truck equivalency factors TF. According to the linear regression given in Fig. 3, an increase in OV from 5 to 23% causes increase in TF from 0.53 to 0.83. The linear relationship does not seem to be very strong and coefficient of determination $R^2 = 0.33$. It is caused by three stations: DK22, A2 and DK79 whose results differ most from the general trend. The relationships between OV and TF analyzed over successive months for individual stations (Rys et al. 2017) yield much stronger correlations (with the coefficient of determination R^2 between 0.76 and 0.99), which indicates that besides overloading of vehicles some local conditions of heavy traffic have effect on truck equivalency factors as well. The model developed by Rys et al. (2016a, b) proved that axle load distribution has an impact on load equivalency factors as well and the model proposed in that publication includes axle load distributions and percentage of overloaded vehicles.

2.4 Decrease in Fatigue Life of Pavement Structures Caused by Overloaded Vehicles

The methodology of determining Decrease in Fatigue Life (DFL) of a pavement structure due to overloaded vehicles was first introduced by Rys et al. (2016a, b, 2017). The factor DFL allows to estimate how increase in percentage of overloaded vehicles contributes to decrease in fatigue life and further decrease in service period of pavement structure. The formula (1) was used to calculate decrease in fatigue life of pavement structures:

$$DFL = 1 - \frac{TF_0}{TF_{OV}} \tag{1}$$

where: DFL—Decrease in Fatigue Life caused by overloaded vehicles, TF_0—Truck equivalency factor at 0% of overloaded vehicles, TF_{OV}—Truck equivalency factors at a given percent of overloaded vehicles OV. The relationship between DFL and percentage of overloaded vehicles OV is shown in Fig. 4 and it is compared with factors DFL obtained from two previous works by the authors (Rys et al. 2016a, b, 2017).

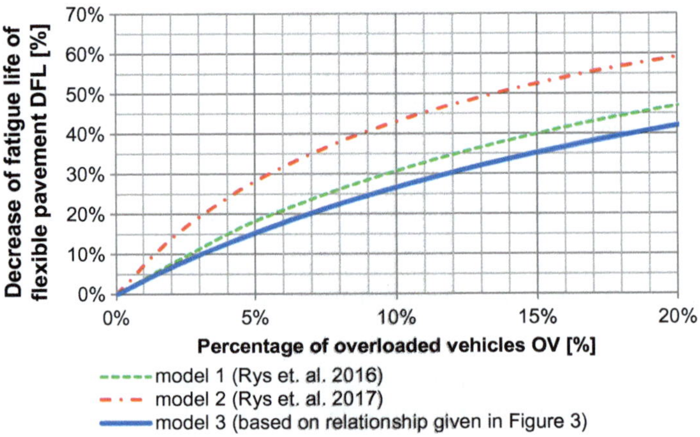

Fig. 4. Relationship between Decrease in Fatigue Life DFL and average percentage of Overloaded Vehicles OV obtained for three different statistical models

It can be concluded from Fig. 4 that three different approaches to determination of impact of percentage of overloaded vehicles on load equivalency factors LEF or truck equivalency factors TF yield similar results of parameter DFL. An increase in percentage of overloaded vehicles from 0 to 20% results in a decrease in fatigue life by 40 up to 60%. For further analysis the model 3 given in Fig. 4 was used to calculate increase in service period due to reduction of the number of overloaded vehicles.

2.5 Extension of Residual Fatigue Life and Service Period of Flexible Pavement Structures Due to Reduction of Number of Overloaded Vehicles

On the basis of Eq. (1) and Fig. 4 the following simple approach is proposed to estimate the effect of reduction in percentage of overloaded vehicles on the extension of residual fatigue life and service period of flexible pavement structure (Rys et al. 2017). The fatigue life depends on the truck equivalency factor TF and total number of trucks NT in the entire service period. Better traffic control may cause reduction in percentage of overloaded vehicles OV thus reducing the value of the truck equivalency factor TF. If the truck equivalency factor TF decreases as a consequence of decrease in the percentage of overloaded vehicles OV, the total number of trucks NT which the pavement can carry before failure will increase and consequently, the residual fatigue life and service period will be extended. To evaluate the extension of residual fatigue life and service period of flexible pavement structures due to decrease in number of overloaded vehicles let us assume that the residual fatigue life of a given pavement was assessed based on measurement obtained from the FWD test. The percentage of overloaded vehicles OV for that road is known. To calculate the residual fatigue life and service period of the pavement in case when number of overloaded vehicles is reduced by ΔOV the following formula can be used:

$$RFL_{OV-\Delta OV} = RFL_{OV}\frac{1 - DFL_{OV-\Delta OV}}{1 - DFL_{OV}} \tag{2}$$

where: $RFL_{OV-\Delta OV}$—Residual Fatigue Life, expressed as number of Equivalent Single Axle Loads, after reduction of overloaded vehicles percentage from OV to (OV − OV), RFL_{OV}—Residual Fatigue Life, expressed as number of Equivalent Single Axle Loads, at the existing number of overloaded vehicles OV, $DFL_{OV-\Delta OV}$—Decrease in Fatigue Life at reduced percentage of overloaded vehicles (OV − ΔOV), DFL_{OV}—Decrease in Fatigue Life at existing percentage of overloaded vehicles OV. The DFL value (Decrease in Fatigue Life) is given in Fig. 4, as a function of percentage of overloaded vehicles OV.

Similarly the extended residual service period of a pavement after reduction of number of overloaded vehicles by ΔOV can be calculated from the formula:

$$RSP_{OV-\Delta OV} = RSP_{OV}\frac{1 - DFL_{OV-\Delta OV}}{1 - DFL_{OV}} \tag{3}$$

where: $RSP_{OV-\Delta OV}$—Residual Service Period, expressed in years, after reduction of overloaded vehicles percentage from OV to (OV − ΔOV), RSP_{OV}—Residual Service Period, expressed in years, at the existing number of overloaded vehicles OV. Both values—Residual Fatigue Life as well as Residual Service Period—will increase by the Increase Factor IF which expresses extension of service period and is equal to:

$$IF = \frac{1 - DFL_{OV-\Delta OV}}{1 - DFL_{OV}} \quad (4)$$

where IF is the Increase Factor, other abbreviations as above. The chart with IF for three target percentages of overloaded vehicles is presented in Fig. 5.

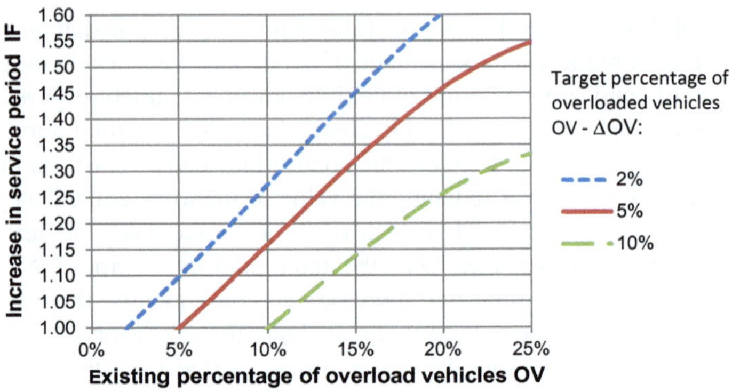

Fig. 5. Increase in service period IF due to reduction of existing percentage of overloaded vehicles to the level of 2, 5 or 10%

The presented method is approximate but may well suit a purpose of evaluating the benefits of traffic control measures. The exact value of increase in fatigue life and service period due to decrease in percentage of overloaded vehicles is dependent on several factors which may vary between specific roads. These factors are: composition and characterization of heavy traffic, current percentage of overloaded vehicles, the degree of overloading, pavement structure and materials used for pavement layers, climatic conditions, etc.

3 Example of Life Cycle Cost Analysis of Pavement with Existing and Reduced Percentage of Overloaded Vehicles

The calculations were performed for an example of national road DK94c with existing percentage of overloaded vehicles equal to 23%, Annual Average Daily Truck Traffic AADT = 1125 trucks per day per lane and truck equivalency factor TF = 0.75 standard 100 kN axles per lane. The typical design period for national roads in Poland equals 20 years. Two alternative scenarios of pavement structure loading were considered:

A. Pavement structure is loaded by heavy traffic with current 23% of overloaded vehicles and service period is equal to design period,
B. The percentage of overloaded vehicles is decreased to 5% and the service period is extended by factor IF = 1.5 (according to Fig. 5).

The same fatigue life of pavement structure was assumed for both alternatives A and B. A plot of pavement condition changes in the following years of service for alternative A and B is presented in Fig. 6. Decrease in OV from 23 to 5% will cause an increase in service period of new structure from designed 20 years to 30 years. Similarly the period between pavement structural rehabilitation will increase from designed 10 years to 15 years.

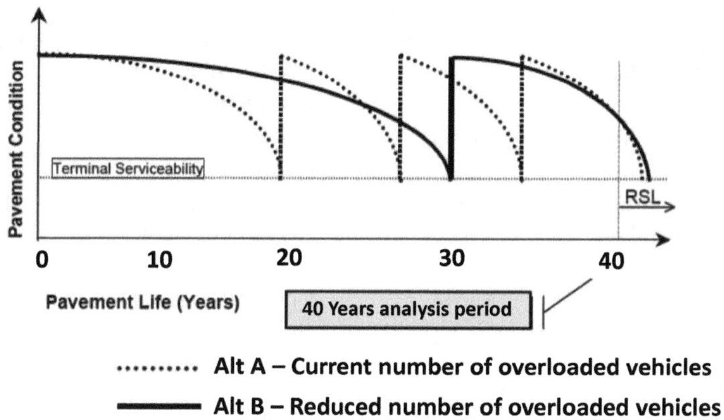

Fig. 6. Effect of reduction of overloaded vehicles on pavement condition and pavement service period

The whole life cost analysis was performed on the basis of instructions (FHWA 1998). The same initial costs and maintenance cost were assumed in both alternatives A and B. The treatment scenario can significantly vary in different countries and it depends on several factors including policy, traffic and climatic conditions, road authority experience etc. The analysis uses an example treatment scenario, which is given in Table 2. The treatment scenario was assumed on the basis of literature review available in work of Lisowska and Ludzik (2016), where a comparison of experience from several countries including USA, Canada, France and Switzerland is given.

It was assumed that costs include discount rate, which equals 2.5% and is constant over the whole period of analysis. More detailed information about maintenance strategy and costs for both alternatives A and B are given in Table 2. The costs given in Table 2 were assumed on the basis of interview with three road construction companies and Polish road authority. In order to make the analysis clearer, only costs bare by road authority were included. The presented analysis does not include user costs but it is expected that consideration of user costs would additionally increase the disproportion between alternatives A and B.

It can be concluded from Table 2 that for the assumed treatment scenario the whole life costs borne by road authority can decrease by 11% if the percentage of overloaded vehicles decreases from 23 to 5%. The results could be slightly different for other treatment scenarios, however the extension of service period will always result in cost

Table 2. Example of maintenance treatment and calculation of costs bare by road authority in case of existing and reduced percentage of overloaded vehicles

Year of analysis	Alternative A Existing percentage of overloaded vehicles OV = 23%		Alternative B Percentage of overloaded vehicles reduced to OV = 5%	
	Maintenance treatment	Cost bare by road authority (per one km)	Maintenance treatment	Cost bare by road authority (per one km)
0	**Initial cost—construction of new pavement**	**$1,342,857**	**Initial cost—construction of new pavement**	**$1,342,857**
5	Crack sealing (~200 m)	$1768	Crack sealing (~200 m)	$1768
10	Crack repairs Thin overlay	$34,523	Crack sealing Repairs of potholes with patches 2% of lane area	$3348
15	Crack sealing Repairs of potholes with patches 2% of lane area	$3348	Crack repairs, milling and reconstruction of wearing course	$47,346
20	Milling of wearing course, overlay with 12 cm of HMA	$217,954	Crack sealing Repairs of potholes with patches 5% of lane area	$6975
25	Crack repairs	$2312	Crack sealing Repairs of potholes with patches 5% of lane area	$6164
30	Crack sealing Repairs of potholes with patches 5% of lane area	$5448	Milling of wearing course, overlay with 8 cm of HMA	$170,265
35	Full-depth reconstruction	$216,705	Crack sealing and repairs	$1806
40	Crack sealing	$745	Crack sealing Repairs of potholes with patches 5% of lane area	$4256
	Salvage value	**−$102,567**		**−$56,782**
	Total cost	$1,723,093		$1,528,004

benefits. The value of 11% does not seem to be very high, however considering maintenance costs of the entire road network it brings multi-million savings.

4 Conclusions

(1) Data from Weigh in Motion stations on seven state roads and one motorway in Poland indicated that the percentage of overloaded vehicles was in the range from 5 to 23%. The lowest percentage was noted on a road where continuous control of traffic was performed.
(2) Most overloaded vehicles exceeded the maximum legal axle load limit, while the gross weight was exceeded less frequently. It indicates that very often freight inside vehicles is not properly distributed which may result from negligence and unawareness of drivers and transport companies.
(3) The relationship between average percentage of overloaded vehicles and average truck equivalency factor was found in the research. According to the linear regression, an increase in percentage of overloaded vehicles from 5 to 23% causes increase in truck equivalency factor from 0.53 to 0.81—by 1.52 times. Increase in truck equivalency factor results in decrease in fatigue life of pavement structure.
(4) Reduction of number of overloaded vehicles causes extension of residual fatigue life and service period of pavement structures. The methodology of estimation of Increase Factor IF which expresses extension of service period was developed in the paper. Reduction of percentage of overloaded vehicles from 23 to 5% causes increase in service period by IF = 1.5.
(5) The example analysis of effect of overloaded vehicles on life cycle cost of pavement structure was performed. For a given example the whole life cycle costs bare by road authority can decrease by 11% when percentage of overloaded vehicles decreases from 23 to 5%. The results may be slightly different for other treatment scenarios, however the extension of service period will always result in cost benefits. Reduction in number of overloaded vehicles can brings multi-million savings when maintenance costs of the entire road network are considered.

References

Al-Qadi, I., Ouyang, Y., Wang, H., Meidani, H., Gungor, O.E., Petit, A., Zhao, J., Qiu, J.: Development of Overweight Vehicle Permit Fee Structure in Illinois. Illinois Center for Transportation, Report FHWA-ICT-17-004 (2017)

American Association of State Highway and Transportation Officials (AASHTO): The AASHTO Guide for Design of Pavement Structures. Washington (1993)

Budzynski, M., Rys, D., Kustra, W.: Selected problems of transport in port towns—tri-city as an example. Pol. Marit. Res. **24**(s1) (2017) https://doi.org/10.1515/pomr-2017-0016

Burnos, P., Gajda, J.: Thermal property analysis of Axle load sensors for weighing vehicles in weigh-in-motion system. Sensors, **16**(12) (2016). https://doi.org/10.3390/s16122143

Burnos, P., Rys, D.: The effect of flexible pavement mechanics on the accuracy of Axle load sensors in vehicle weigh-in-motion systems. Sensors, **17**(9) (2017). https://doi.org/10.3390/s17092053

Burnos, P., Gajda, J., Piwowar, P., Sroka, R., Stencel, M., Zeglen, T.: Accurate of weighing of moving vehicles. Metrolo. Meas. Syst. **14**(4), 507–516 (2007)

COST 323: Weigh-in-Motion of Road Vehicles. Final Report. Appendix 1. European WIM Specifications. LCPC, Paris (2002)

Dey, K., Chowdhury, M., Wiecek, M., Dunning, A.: Infrastructure damage-cost-recovery fee for overweight trucks: tradeoff analysis framework. J. Transp. Eng. (2015). https://doi.org/10.1061/(asce)te.1943-5436.0000769,04015008

Doupal, E., Calderara, R.: Combined LS & HS WIM system for law enforcement and toll road applications. In: Proceedings of the 10th International Conference on Heavy Vehicle, France, 2008

Federal Highway Administration (FHWA): Life-cycle cost analysis in pavement design—interim technical bulletin. FHWA-SA-98-079. Washington (1998)

Federal Highway Administration (FHWA): WIM Data Analyst's Manual. Publication No. FHWA-37 IF-10-018. Washington (2010)

Fiorillo, G., Ghosn, M.: Procedure for statistical categorization of overweight vehicles in a WIM database. J. Transp. Eng. **140**(5) (2014). https://doi.org/10.1061/(asce)te.1943-5436.0000655

General Directorate for National Roads and Motorways (GDDKiA): The Synthesis of the General Traffic Measurement. Warsaw (2015). (Available in the Internet: www.gddkia.gov.pl/pl/987/gpr-2015)

Jacob, B. Loo, H.: Weight in motion for enforcement in Europe. In: 10th International Symposium on Heavy Vehicle Transportation Technology HVTT10, France, 2008

Jeongho Oh, A..M, Fernando, E.G., Lytton, R.L.: Evaluation of damage potential for pavements due to overweight truck traffic. J. Transp. Eng. **133** (2007). https://doi.org/10.1061/(asce)0733-947x(2007)133:5(308)

Judycki, J.: Determination of equivalent Axle load factors on the basis of fatigue criteria for flexible and semi-rigid pavements. Road Materi. Pavement Des. **11**(2010), 187–202 (2010). https://doi.org/10.1080/14680629.2010.9690266

Judycki, J.: Bases for determination of the equivalent axle load factors for design of road design of road pavements. Drogi i Mosty (Road and Bridges) no 2/2006 (2006) (in Polish)

Judycki, J., Jaskuła, P., Pszczoła, M., Ryś, D., Jaczewski, M., Alenowicz, J., Dołżycki, B., Stienss, M.: New polish catalogue of typical flexible and semi-rigid pavements. In: XI International Road Safety Seminar GAMBIT 2016, MATEC Web of Conferences, vol 122, 2017. https://doi.org/10.1051/matecconf/201712204002

Lisowska, K., Ludzik, D: A comparative analysis of construction and maintenance costs of asphalt and cement concrete expressway, M Sc thesis, Gdansk University of Technology (2016) (in Polish)

Mohammadi, J., Shah, N.: Statistical evaluation of truck overloads. Journal of Transportation Engineering, **118**, 651–665. ASCE (1992). https://doi.org/10.1061/(asce)0733-947x(1992)118:5(651)

Mulyono, A.T., Antameng, M. Analysis of loss cost of road pavement distress due to overloading freight transportation. J. Eastern Asia Soc. Transp. Stud. **8** (2010)

National Cooperative Highway Research Program NCHRP: Traffic data collection, analysis, and forecasting for mechanistic pavement design. Report 538 (2005)

Oskarbski, J., Kaszubowski, D.: Implementation of weigh-in-motion system in freight traffic management in urban areas. In: 2nd International Conference Green Cities—Green Logistics for Greener Cities, Transportation Research Procedia, 2016, vol. 16, pp. 449–463. https://doi.org/10.1016/j.trpro.2016.11.042

Pais, J.C., Amorim, S.I.R., Minhoto, M.J.C.: Impact of traffic overload on road pavement performance. J. Transp. Eng. **139**(9), 873–879 (2013). https://doi.org/10.1061/(ASCE)TE.1943-5436.0000571

Pszczoła, M., Judycki, J., Ryś, D.: Evaluation of pavement temperatures in Poland during winter conditions. Transport research arena TRA2016. Transp. Res. Proc. **14**, 738–747 (2016). https://doi.org/10.1016/j.trpro.2016.05.342

Rys, D., Judycki, J., Jaskula, P.: Analysis of effect of overloaded vehicles on fatigue life of flexible pavements based on weigh in motion (WIM) data. Intern. J. Pavement Eng. **17**(8) (2016a). https://doi.org/10.1080/10298436.2015.1019493

Rys, D., Judycki, J., Jaskula, P.: Determination of vehicles load equivalency factors for polish catalogue of typical flexible and semi-rigid pavement structures. Transport research arena TRA2016. Transp. Res. Proc. **14**, 2382–2391 (2016b). https://doi.org/10.1016/j.trpro.2016.05.272

Rys, D., Judycki, J., Jaskula, P.: Impact of overloaded vehicles on load equivalency factors and service period of flexible pavements. In: Proceedings of the 10th International Conference on the Bearing Capacity of Roads, Railways and Airfields (BCRRA 2017), 28–30 June 2017, Athens, Greece. ISBN 9781138295957. https://doi.org/10.1201/9781315100333

Stephens, J., Carson, J., Hult, D.A., Bisom, D.: Preservation of infrastructure by using weight-in-motion coordinated weight enforcement. Transp. Res. Rec. 1855, 143–150. TRB (2003). https://doi.org/10.3141/1855-18

Szydlo, A., Wardega R.: The comparison of the vehicles aggressiveness factors in Poland and in the European Union. Drogownictwo no 5/2003 (2003) (in Polish)

Tailor, B., Bergan, A., Lindgren, N., Berthelot, C.: The Importance of commercial vehicle weight enforcement in safety and road asset management. Traffic Technol. Intern. (2000)

Wang, H., Zhao, J.: Development of overweight permit fee using mechanistic-empirical pavement design and life-cycle cost analysis. Transport **31**(2), 156–166 (2016). https://doi.org/10.3846/16484142.2016.1191039

Zhao, Y., Tan, Y., Zhou, C.: Determination of axle load spectra based on percentage of overloaded trucks for mechanistic-empirical pavement design. J. Road Mater. Pavement Des. **13**(4), 850–863 (2012). https://doi.org/10.1080/14680629.2012.735796

Zofka, A., Urbaniak, A., Maliszewski, M., Bankowski, W., Sybilski, D.: Site specific traffic inputs for mechanistic-empirical pavement design guide in Poland. In: Transportation Research Board, 93rd Annual Meeting, Washington, USA, 2014

Practical Pavement Design Approach Subjected to Seepage Conditions

Anand Tapase[1(✉)] and M. S. Ranadive[2]

[1] Department of Civil Engineering, Rayat Shikshan Sanstha's, Karmaveer Bhaurao Patil College of Engineering, Satara, Maharashtra, India
tapaseanand@gmail.com

[2] Department of Civil Engineering, College of Engineering, Pune, Maharashtra, India
msr.civil@coep.ac.in

Abstract. The water seeps through the base and subgrade which results in a decrease in the life of the pavement by a reduction in modulus of underneath layers. Accurate analysis of water seepage through soil is essential to achieve a more durable design of pavement. A present-day system demands an application of analytical tool which will enhance the scope to find the unknown nodal values along with governing parameters like seepage pressures and gradients. In this connection, the advantages of the versatile finite element technique over the other available methods for seepage analysis holds a bright promise. This study illustrates the effectiveness of finite element analysis of two-dimensional steady-state seepage fields with the confined flow boundaries in considering a practical design approach for achieving the optimum pavement section subjected to seepage conditions. In the analysis, it is assumed that the energy supplied to the soil mass is entirely dissipated due to the seepage through the mass, which provides the basis for the formulation of the set of governing equations. The solution gives the unknown nodal discharge. It is noticed that the discharge at the nodes representing the inflow into the medium is positive, and is negative for the nodes representing the outflow from the medium. Such type of analysis proves beneficial for deriving useful design charts/procedures by correlating the obtained data from analysis with the actual field condition.

1 Introduction

Development of any country depends on its road network which contributes in exchanging the social, cultural, economic and industrial activities; on other hands the increasing traffic along with overloaded vehicles, actual field conditions are responsible for early development of distress symptoms like raveling, undulations, rutting, cracking, bleeding and potholing in bituminous layer (Ranadive and Tapase 2013; Tapase and Ranadive 2017). The water seeps through the base and subgrade which results in the decrease in the life of the pavement by a reduction in modulus of underneath layers. Accurate analysis of water seepage through soil is essential to achieve a more durable design of pavement. With a view to finding the unknown nodal values along with governing parameters in the present work, two-dimensional steady-state seepage

K. Zhang et al. (eds.), *Testing and Characterization of Asphalt Materials and Pavement Structures*, Sustainable Civil Infrastructures,
https://doi.org/10.1007/978-3-319-95789-0_11

analysis is performed considering the fields with the confined flow boundaries. A present-day system demands an application of analytical tool which will enhance the scope to find the unknown nodal values along with governing parameters like seepage pressures and gradients. In this connection, the advantages of the versatile finite element technique over the other available methods for seepage analysis holds a bright promise.

2 Literature Review

Accurate analysis of water seepage through soil is essential to achieve the more durable design of pavement. The seepage is a very important phenomenon to be reckoned with while designing the soil engineering systems, wherein the seepage through the soil mass governs their behavior. Various methods are successfully employed to analyze the seepage problems.

Amongst various techniques, the graphical method of sketching the flow nets offers the simplest method for the solution of seepage problems. Even though a soil mass may appear to be uniform and homogeneous, its coefficient of permeability may erratically vary with depth. In general, the soil masses display anisotropy with respect to the coefficient of permeability. So long as the anisotropic ratio and the direction of anisotropy remain same throughout the seepage medium, the graphical method may easily be employed after suitable geometrical transformation of the medium. However, more often than not, the ratio, as well as the orientation of anisotropy, may vary so much from point to point that even the sophisticated schemes may fail in affecting the geometrical transformation needed for applying the graphical method. In view of the above, it is concluded that in case of the highly heterogeneous soil masses, it is impossible to achieve the solution by employing the graphical method.

Where the graphical method fails, it is impossible to provide a rigorous mathematical solution. Even though the mathematical treatment has the capacity to provide a most accurate solution in closed form, it can handle only simple seepage media and the even geometrical boundaries. Obviously, the method has limitations in offering solution to the seepage problems, associated with heterogeneous soil masses or the medium having uneven geometrical boundaries.

Analog solutions take the advantages of the common governing equation, even though the fields may be totally diverse. For example, the standard Laplace equation, which governs the steady state seepage through the isotropic homogeneous soil media, also governs the phenomena, such as the flow of current through electrical conducting system deflections of a thin membrane, etc. Taking the advantage of this fact, the seepage problems could be solved by employing the techniques of the electrical analogy, the membrane analogy, etc.

The analog solutions are superior to mathematical solution due to their ability to provide satisfactory results in those cases where the mathematics would fail. In addition, they are more accurate in comparison with the solution derived through the graphical methods. In the present state of the developments in the technique of analog solution, it appears that the analog solution can accommodate sufficient degree of heterogeneity of the medium, though they may fail in case of highly heterogeneous seepage media.

Numerical solutions do not have the above kind of limitations. Even though the numerical methods offer only approximate solutions, it is possible to achieve realistic results due to the availability of the high-speed digital computers. Amongst the solutions, those based on the principles of finite differences became most popular since 1940. The finite difference method has the capacity to accommodate complex geometrical boundaries as well as the highest degree of heterogeneity into the analysis. The finite element method has also this capacity. In addition, Zienkiewicz and Taylor (1991) states that the finite element method, when applied to the seepage problems, has the following advantage over the method of finite differences.

(a) The simplicity of dealing with non-homogeneous and anisotropic situations (Particularly when the direction of anisotropy is variable).
(b) The elements can be graded in shape and size to follow arbitrary boundaries and to allow for the regions of rapid variation of the flow potential.
(c) Specific gradients are prescribed by the stream boundary conditions which are introduced naturally and with better accuracy than the finite difference procedures.
(d) Higher order elements can be used to improve accuracy without complicating boundary conditions, a difficulty always arising with finite difference approximation of the higher order.
(e) Finally but of considerable importance, in the computer age, standard programs may be used for assembly and solution.

Kalyani and Gunarate (2009) developed a model to analyze the effects of steady-state seepage of water based on continuum models. A water pressure gradient is applied across one two-layer unit to trigger water seepage and the pore water motion is idealized using Navier–Stokes (NS) equations.

3 Pavement Section and Finite Element Analysis

The two-dimensional finite element method for the analysis of the steady state seepage fields is performed. The scope of the present work is limited to the seepage fields with the confined flow boundaries.

Engineering problems are often governed by a large number of parameters and complexities, which are represented by a set of equations, whereas the source programs are prepared which on its execution results in a solution to those unknown entities. Constitutive models represent a mathematical model that describes the behavior of engineering materials play a significant role in providing reliable results from any solution procedures. Increasing use and development of computer-based techniques such as finite difference method, boundary integral equation method, and the finite element method have enhanced the importance of constitutive models of engineering materials. The physical behavior of the engineering problem once understood is established in the form of a mathematical model for solving it quantitatively. The internal mechanical response of a material can be expressed in terms of stresses and

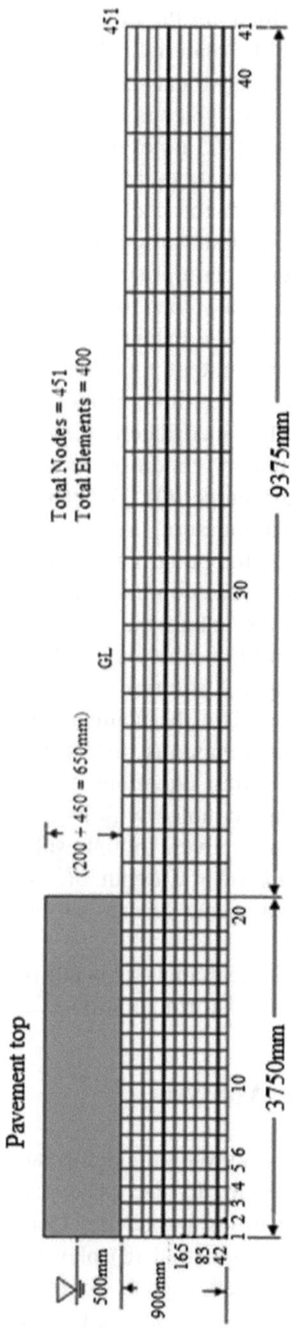

Fig. 1. Idealization for seepage analysis

strains. The constitutive laws which are the outcome of extensive laboratory and field test are applied herein to define the material behavior.

In the present study, a two-dimensional seepage analysis is performed, wherein the thickness of the subgrade is 900 mm which is kept constant throughout the analysis and the lateral extent is assumed to be at 2.5 times the pavement width which represents the indefinite lateral extent of the soil strata adequately. A system comprising of the two-layer pavement of total 650 mm thickness comprising of 450 mm base layer and 200 mm of BL resting over soil strata/subgrade is analyzed (Fig. 1). In the present study, a coarse mesh for idealization comprising of 400 quadrilateral elements and 451 nodes is employed. Lateral extent is assumed to be at 2.5 times the pavement width from the toe of the base layer (i.e. horizontal extent is taken as 3750 + 9375 = 13,125 mm) represents the indefinite lateral extent of the soil strata adequately. For the two-dimensional seepage analysis, the finite element idealization for the pavement is developed by means of four noded quadrilateral elements (Fig. 1).

It is recognized that the boundary conditions comprise of the prescribed values of nodal flow potential or nodal discharge (Q). For example, at the nodes lying over the discharge face of the atmospheric pressure, the values of flow potential are known, but the values of Q are unknown. On the other hand, at the remaining nodes, inflow being equal to outflow due to the steady state condition, the values of Q are zero, but the flow potential values at those nodes are unknown.

In the present case, an algorithm wherein the logical rules including the mathematical expressions representing the solution of the problem in the form of critical parameters is interlinked to formulate the source program. The developed computer source program executes the unambiguous orders that are logically interlinked with mathematical expressions, governing equations along with boundary conditions in terms of the FORTRAN language coupled with its operations, which interns provides the results at the critical location. Even though subgrade soils are not elastic and undergo deformation, depending on the depth of the point considered, the vertical stresses at certain depth yields the same stresses for both linear and nonlinear theories (Huang 2008).

The accuracy of the finite element analysis is affirmed by comparing the computed results with readily available closed-form solution (Tapase and Ranadive 2016).

4 Steady State Seepage Analysis

There is a good deal of similarity between the deformation analysis and the steady state seepage analysis considered herein. In the steady state seepage analysis, the flow potential ($\emptyset$) represents the governing parameter. Consequently, in the finite element seepage analysis, the nodal flow potential ($\emptyset$) play the role analogous to that of the nodal displacements in the deformation analysis. In a similar manner, the nodal discharge (Q) in the seepage analysis is analogous to that of the nodal forces in the deformation analysis.

Theoretically, a soil mass subjected to seepage would deform due to the body forces induced by the seepage. In case of the steady state seepage analysis, absence of such deformations, (i.e., the soil mass is treated as a rigid porous body) is assumed. As a consequence of this, it is assumed that the energy supplied to the soil mass is entirely dissipated due to the seepage through the mass. This would provide the basis for the formulation of the set of governing equations.

5 Gradients and Velocities

Let i_x and i_y be the potential gradients in the 'x' and the 'y' directions respectively. Let these gradients be represented through the vector $[i]$ then,

$$[i] = \frac{i_x}{i_y} = \frac{\frac{\partial \emptyset}{\partial x}}{\frac{\partial \emptyset}{\partial y}} = [\mathbf{B}][\emptyset_\mathrm{e}] \tag{1}$$

wherein, (B) matrix is composed of the constant coefficients, hence, throughout the element the gradients are constant. This is analogues to the plane strain element.

Let, v_x and v_y be the flow velocities in the x and y directions respectively; then according to the well known Darcy's law;

$$\begin{aligned} v_x &= -k_x i_x \\ v_y &= -k_y i_y \end{aligned} \tag{2}$$

The minus sign in the Eq. 2 is due to the fact that in the positive direction of the velocities, the flow potential ($\emptyset$) should decrease; hence the gradient in those directions would be negative. These components of the velocities are represented through a vector $[v]$, consequently,

$$[v] = \frac{v_x}{v_y} = [\mathrm{c}][i] = -[\mathrm{C}][\mathrm{B}][i] \tag{3}$$

Like the gradients, the velocities are also constant throughout the element. The gradients and the velocities for the elements are computed by means of the Eqs. 2 and 3 respectively. The nodal gradients and the nodal velocities are then derived by the method of direct averaging.

6 Seepage Characteristics

The seepage forces of intensities "p_x" and "p_y" per unit volume is acted in 'x' and 'y' directions respectively; wherein;

$$\begin{aligned} p_x &= -\gamma_w i_x \\ p_y &= -\gamma_w i_y \end{aligned} \tag{4}$$

So that γ_w is the unit weight of water.

The minus sign in the Eq. 5 is due to the fact that in the positive direction of the seepage forces, the flow potential (∅) decreases, hence the gradients are negative in those directions.

The vector (P) of the seepage forces, is as defined in the Eq. 5.

$$[\mathrm{P}] = -\gamma_w \frac{i_x}{i_y} = -\gamma_w[i] = -\gamma_w[\mathrm{B}][\emptyset_\mathrm{e}] \quad (5)$$

As (i_x, i_y) are constant throughout the element, the same is also the case with (p_x, p_x).

For the steady state seepage condition,

$$[Q_e] = [\mathrm{S_e}][\emptyset_e] \quad (6)$$

Wherein, [Se] is the element seepage matrix as defined in the Eq. 7

$$[\mathrm{S_e}] = \Delta \cdot [B]^T[c][B] \quad (7)$$

and;

$$[\mathrm{Q}] = [\mathrm{S}][\phi] \quad (8)$$

Wherein; [S] is the final seepage matrix, which could be derived by super positioning of [Se] of all the ‘N’ elements;

7 Boundary Conditions

It is recognized that the boundary conditions comprises of the prescribed values of nodal flow potential (ϕ) or nodal discharge (Q), for example at the nodes lying over the discharge face at the atmospheric pressure, the values of flow potential (ϕ) are known, but the values of nodal discharge (Q) are unknown. On the other hand at the remaining nodes, in general inflow being equal to outflow due to the steady state condition, the nodal value of discharge (Q) is zero, but the flow potential (ϕ) values at those nodes are unknown. It is recognized that the boundary conditions would comprise of the prescribed values of nodal ‘ϕ’ or nodal ‘Q’. For example at the nodes lying over an equipotential surface or the discharge faces at the discharge faces at the atmospheric pressure, the values of ‘ϕ’ are known, but the values of ϕ are unknown. on the other hand at the remaining nodes, in general inflow being equal to outflow due to the steady state condition,the nodal values of ‘Q’ are Zero, but the ‘ϕ’ values at those node are unknown. Representing the knowns and the unknowns through the suffixes ‘K’ and ‘u’ respectively, the equation is arranged as given in Eq. 8.

$$\frac{[S_I]}{[S_{II}]^T}\ \frac{[S_{II}]}{[S_{III}]}\ \frac{[\phi_U]}{[\phi_K]} = \frac{[Q_K]}{[Q_U]}$$

Therefore, (11)

$$[S_I][\phi_u] + [S_{II}][\phi_K] = [Q_K]$$
$$[S_{II}][\phi_U] + [S_{III}][\phi_K] = [Q_U]$$

From the first part of the Eq. 8, $[\phi_u]$ is obtained and on substituting it in the second equation $[Q_U]$ is derived. Thus the unknown nodal (ϕ, Q) gets established with the help of these the gradients and velocities developed in the medium could be derived.

In the analysis following boundary conditions are employed,

(i) From Fig. 1, the nodes 1, 42, 83, 124, 165, 206, 247, 288, 329, 370, 411 are at entrance so the values of discharge (Q) are unknown and the values of potential flow are known which is taken as 100%.
(ii) At the nodes open to atmosphere, where the value of discharge (Q) is unknown, flow potential/pressure ϕ is taken to be equal to zero.
(iii) At the remaining nodes discharge (Q) is equal to zero and value of flow potential (ϕ) is unknown.

8 Results and Interpretation

This study illustrates the effectiveness of finite element analysis of two-dimensional steady-state seepage fields with the confined flow boundaries in considering a practical design approach for achieving the optimum pavement section subjected to seepage conditions. In the analysis, it is assumed that the energy supplied to the soil mass is entirely dissipated due to the seepage through the mass, which provides the basis for the formulation of the set of governing equations. The solution gives the unknown nodal discharge (Q) as mentioned in Table 1. It is noticed the discharge (Q) at the nodes 1, 42, 83, 124, 124, 165, 206, 247, 288, 329, 370 and 411 are positive; this is because they represent the inflow into the medium. On the other hand, negative discharges indicate outflow from the medium. In the steady state seepage condition, inflow is equal to outflow; the slight difference between inflow and outflow is due to the rounding off of the numerical computations at the various stages. Let the 100% head be equal to 500 mm head of water, with this the percentage nodal values of flow potential (ϕ) are transformed into the equivalent head of water. A two-dimensional steady-state seepage analysis with the confined flow boundary is performed, wherein the deformation due to the soil mass body forces induced by the seepage is assumed to be absent (i.e. the soil mass is treated as rigid porous). The results pertaining to the nodes of flow value in percentage and equivalent meter head at nodes are presented in Table 1.

Table 1. Value of unknown nodal discharge

Node number	Flow potential (ϕ)	Discharge (Q)
1	1.00E+02	1.01E+00
42	1.00E+02	2.01E+00
83	1.00E+02	2.01E+00
124	1.00E+02	2.01E+00
165	1.00E+02	2.01E+00
206	1.00E+02	2.01E+00
247	1.00E+02	2.01E+00
288	1.00E+02	2.01E+00
329	1.00E+02	2.01E+00
370	1.00E+02	2.01E+00
411	1.00E+02	1.01E+00
432	0.00E+00	−1.07E+01
433	0.00E+00	−5.05E+00
434	0.00E+00	−2.15E+00
435	0.00E+00	−1.11E+00
436	0.00E+00	−5.62E−01
437	0.00E+00	−2.89E−01
438	0.00E+00	−1.48E−01
439	0.00E+00	−7.59E−02
440	0.00E+00	−3.89E−02
441	0.00E+00	−2.30E−02
442	0.00E+00	−1.18E−02
443	0.00E+00	−4.21E−03
444	0.00E+00	−1.51E−03
445	0.00E+00	−5.40E−04
446	0.00E+00	−1.93E−04
447	0.00E+00	−6.92E−05
448	0.00E+00	−2.48E−05
449	0.00E+00	−9.01E−06
450	0.00E+00	−3.58E−06
451	0.00E+00	−1.14E−06

9 Conclusions

Based on the work undertaken following are some of the important conclusions which are directly or indirectly applicable in practice. Considering 100% as head of water for various trial heights, the percentage nodal values of flow potential are transformed into the equivalent head of water. The solution gives the unknown nodal discharge. It is noticed that the discharge at the nodes representing the inflow into the medium is positive, and is negative for the nodes representing the outflow from the medium. Such type of analysis proves beneficial for deriving useful design charts/procedures by correlating the obtained data from analysis with the actual field condition.

References

American Association of State Highway Officials (AASHTO): AASHTO guide for the design of pavement structures. AASHTO, Washington, D.C. (1993)

Huang HY (2008) Pavement analysis and design, 2nd edn. Pearson Education, Inc, and Dorling Kindersley Publishing, Inc., New York

IRC: 37-2012. Guidelines for the design of flexible pavements. Indian Roads Congress, New Delhi

Kalyani, J., Gunarate, M.: Analysis of water seepage in a pavement system using the particulate approach. Comput. Geotech. **36**(4), 641–654 (2009). https://doi.org/10.1016/j.compgeo.2008.09.002

Ranadive, M.S., Tapase, A.B. (2016a). Parameter sensitive analysis of flexible pavement. Int. J. Pavement Res. Technol. (IJPRT), Elsevier, Special Issue on Sustainability on Pavement Engineering. https://doi.org/10.1016/j.ijprt.2016.12.001

Ranadive, M.S., Tapase, A.B.: Investigation of behavioral aspects of flexible pavement under various conditions by finite element method. In: Yang, Q., Zhang, J.-M., Zheng, H., Yao, Y. (eds.) Constitutive Modeling of Geomaterials, pp. 765–770. Springer, Berlin (2013). https://doi.org/10.1007/978-3-642-32814-5_100

Ranadive, M.S., Tapase, A.: Pavement performance evaluation for different combinations of temperature conditions and bituminous mixes. Innovative Infrastruct. Solutions **1**(40), 15 (2016b). https://doi.org/10.1007/s41062-016-0040-9

Tapase, A., Ranadive, M.: Performance evaluation of flexible pavement using finite element method. In: ASCE GSP 266, GeoChina 2016: material, design, construction, maintenance, and testing of pavement, pp. 9–17. https://doi.org/10.1061/9780784480090.002

Tapase, A., Ranadive, M.S.: Predicting performance of flexible pavement using finite element method. Advancement in the design and performance of sustainable asphalt pavements. In: GeoMEast 2017. Sustainable civil infrastructures. Springer, Cham (2017). https://doi.org/10.1007/978-3-319-61908-8-11

Zienkiewicz, O.C., Taylor, R.L.: The finite element method, vol. 2. McGraw-Hill, New York (1991)

Study on Shrinkage of Ordinary Concrete Under Different Temperatures and Humidity

Jinjin Shi(✉), Shitao Liu, Tongxin Liang, Hongling Yu, and Ying Zhang

Cangzhou Municipal Engineering Company Limited, Cangzhou, China
czszjsk@126.com

Abstract. All cement concrete undergoes drying shrinkage or volume change as the concrete ages. The volume change in concrete is very important to the engineer in the design of a structure. Shrinkage cracks can impact the durability and safety of the structures. The degree of dry shrinkage is affected by many factors. This paper investigated three different cement grades (C30, C40, C50) and three different humidity (30, 60, 95%) at two temperature conditions (10, 25 °C). The dry shrinkage of three cement grades was tested and the results were measured at 7, 14, 28, and 60 days. Through the analysis of the test results, it was found that the humidity has the most profound impact on dry shrinkage. When concrete was subjected to a dry atmosphere, it would lead to a greater drying shrinkage. Lower temperatures generally produce a decrease in drying shrinkage because of higher humidity and slower evaporation.

1 Introduction

Dry shrinkage is an inherent characteristic of cement-based concrete which is the shrinkage of concrete in an unsaturated environment during cement hydration. Drying shrinkage happens mostly because of the reduction of capillary water by evaporation and the water in the cement paste. Concrete is in flowing state during placement and in solid state after hardening. Hardening is a chemical process that produces solid materials such as calcium silicate. It is also a physical process that is accompanied by evaporation of water. When the water evaporation rate is greater than the rate of bleeding, the surface begins to shrink, but it can adapt to the volume change without cracking because of the plastic state of concrete at this time. After that, the concrete becomes thicker because cement hydration and the possibility of plastic cracking increases with the evaporation. After the concrete is set, the drying shrinkage of the hardened concrete may lead to cracking when the amount of shrinkage is large. The water loss rate of concrete is related to many factors, such as water glue ratio, total amount of gelation material, dosage of mineral admixture, temperature and humidity, etc. Ruimin Xiao, Xiong Zhang and Jialin Le have studied that water glue ratio is less affected in the early stage of concrete drying shrinkage and there is a critical value in the later stage. When water glue ratio exceeds concrete critical value, the relationship between the drying shrinkage and the water gel ratio was not obvious. When water glue ratio is below concrete critical value, the concrete drying shrinkage increased with the water glue ratio increase in the later stage. The total amount of gelled material has little

K. Zhang et al. (eds.), *Testing and Characterization of Asphalt Materials and Pavement Structures*, Sustainable Civil Infrastructures,
https://doi.org/10.1007/978-3-319-95789-0_12

effect on the early drying shrinkage of concrete, and the shrinkage value increases with the total amount of gelled material increase [1]. Now concrete is generally mixed with mineral admixtures. For high performance concrete, mineral admixtures are necessary. Different mineral admixture is studied by Kang Bai, Hongfa Yu, etc. the influence of different dosage on drying shrinkage of concrete, the drying shrinkage development within 7 day is rapid, 7–60 day drying shrinkage speed slow, stabilized after 60 day, and they put forward the measure of using expansive agent to reduce drying shrinkage [2]. The effect of temperature and humidity on the shrinkage of concrete is studied in this paper.

2 Determination of Concrete Shrinkage Test Method

There are many concrete shrinkage tests in the world. In China alone, there are China national standard GB/T50082-2009, China Ministry of Communications Standard JTJ270-98 and the China power industry standard DL/T5120-2001. Many international institutions have provided concrete shrinkage deformation test standards, such as ASTM, the European EN, the British BS, Japan JIS. There are differences among the test methods. For example, GB/T50082-2009 included both contact and non-contact methods. The contact method included a concrete shrinkage measuring device and the non-contact method comprised a non-contact concrete shrinkage deformation tester. ASTM C157 and British EN utilized gauge for measuring shrinkage. European EN uses contact sensors to measure shrinkage [3].

The non-contact method of GB/T50082-2009 (China national standard) is mainly testing the shrinkage rate of concrete for the first 3 days [4, 5]. In contrast, the contact method is mainly measuring the dry shrinkage of concrete after 3 days. Thus, the contact method was selected and employed in the study.

3 Preparation of Concrete Specimens

Concrete strength and workability (typically reflected in slump) are the two main controlling parameters in China. The cement grades included in the study are C30, C40 and C50. The slump was controlled at 200 ± 20 mm.

The selected materials were tested to comply with the China national and industry standards.

The specimens were prepared according to the mixing design given in Table 1.

Table 1. Mixing design

Strength	Cement	Mineral powder	Fly ash	Sand	5–10 gravel	10–20 gravel	Water	Water-reducing agent
C30	220	80	80	735	165	937	175	8.4
C40	275	80	80	678	166	941	170	9.6
C50	330	80	80	650	166	941	160	12.3

The size of the shrinkage test specimen/prism was 100 mm × 100 mm × 515 mm. Six sets of specimens were prepared and tested under different temperatures and humidity.

4 Testing

In order to ensure the accuracy and reliability of the test results, the temperatures were controlled at between 0 and 25 °C with accuracy of ±1 °C and humidity was controlled at between 30 and 95% with accuracy ±1%. Three different concrete grades were investigated: (1) C30, (2) C40, and (3) C50. Six specimens were prepared for each concrete grade. Test variables included 2 temperatures (10 and 25 °C) and 3 humidity (30, 60, and 95%).

5 Results and Analysis

The specimens were cured for 3 days in a standard curing environment. Then, they were taken out for 4 h at a controlled temperature of 20 ± 2 °C. Measurements were made to determine their initial lengths. The position and direction of specimens on the contraction should be consistent at all times. The corresponding direction was indicated on the specimen. Test specimens/prisms were placed in the corresponding wet and dry box. When the curing was completed, the length of test specimen was measured, and the concrete shrinkage rate was calculated using Eq. 1:

$$\varepsilon_{st} = (L_0 - L_t)/L_b \tag{1}$$

where ε_{st}—the contraction rate of concrete at test age t, and t is the time when the concrete was stirred with water; L_0—initial reading of specimen length, mm; L_t—the length of the specimen at age t, mm; L_b—the length of the concrete specimen (excluding the protruding part of the side head) minus two of the head was buried in the depth [5].

The test results of shrinkage rates for different curing ages and humidity at curing temperature of 10 are shown in Tables 2, 3, 4 and Figs. 1, 2, 3.

Table 2. Shrinkage rate at 30% humidity

Grade	Age			
	7 days	14 days	28 days	60 days
C30	389	530	815	860
C40	352	552	870	904
C50	322	578	850	887

Units 10^{-6}

Table 3. Shrinkage rate at 60% humidity

Strength grade	Age			
	7 days	14 days	28 days	60 days
C30	91	167	318	382
C40	122	190	325	400
C50	145	210	340	421

Units 10^{-6}

Table 4. Shrinkage rate at 95% humidity

Strength grade	Age			
	7 days	14 days	28 days	60 days
C30	65	87	115	130
C40	109	126	150	168
C50	115	132	160	179

Units 10^{-6}

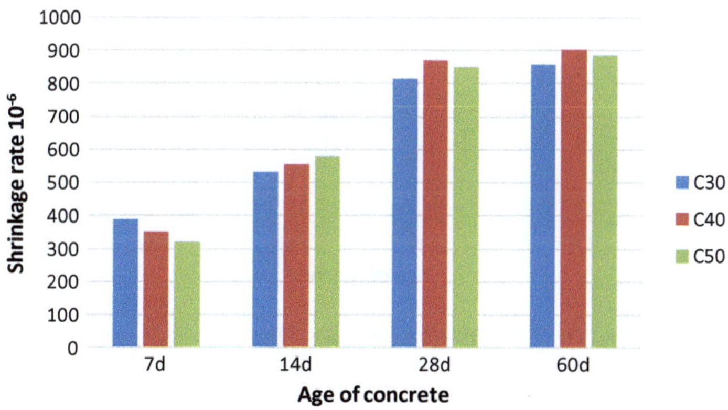

Fig. 1. Shrinkage rate of concrete for curing temperature 10 °C and humidity 30%

Through the analyses of the test results, it was found that shrinkage rate increased with decreasing humidity. It means the shrinkage rate was smaller under higher humidity. This phenomenon was more pronounce for high strength grade than the low strength grade concrete. As expected, the shrinkage rate slowed down after 28 days of curing.

The decrease of the humidity of the concrete itself was the result of the chemical reaction. After the concrete was harden, the volume shrinkage caused by the hydration of the cementitious material was expressed in macroscopic volume shrinkage. In the early stage, there was sufficient internal moisture, so the internal humidity was close to 100%. With time, the rate of hydration decreases and the humidity of concrete decreases as well. When the internal humidity of concrete began to fall, it would cause

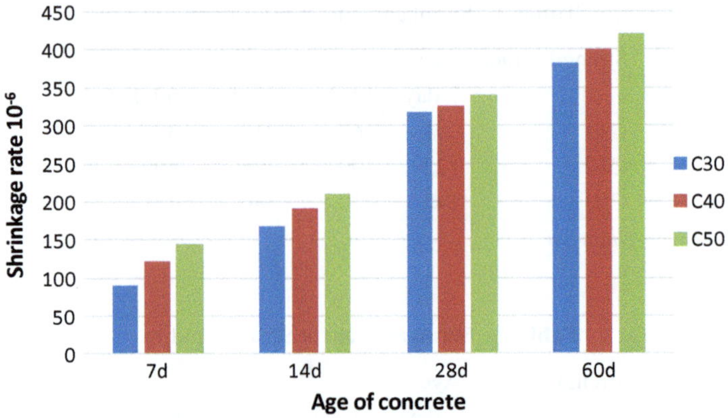

Fig. 2. Shrinkage rate of concrete for curing temperature 10 °C and humidity 60%

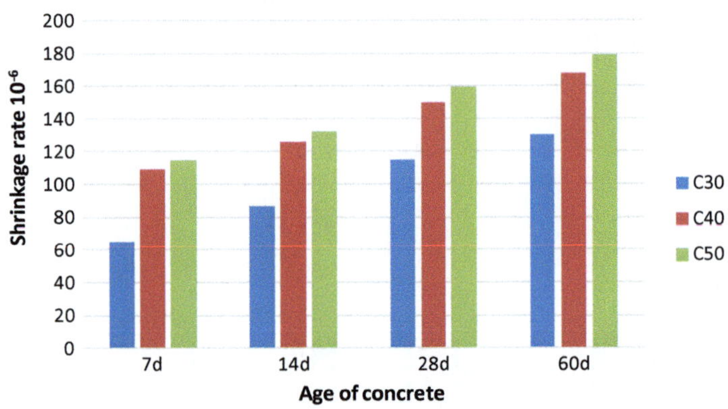

Fig. 3. Shrinkage rate of concrete for curing temperature 10 °C and humidity 95%

the macroscopic shrinkage of concrete, which was controlled by the chemical shrinkage before the internal humidity drops. Therefore, the early shrinkage of concrete was closely related to the change of internal humidity. The internal humidity was reduced because the hydration reaction consumes water. When decrease of humidity in concrete was due to the dissipation of water diffusion, it is called drying shrinkage. Therefore, the change of the internal humidity of concrete was the main driving force of early shrinkage [6].

The test results of shrinkage rate for different cement grade and humidity at curing temperature of 25 °C are shown in Tables 5, 6, 7 and Figs. 4, 5, 6.

In view of the aforementioned test results, it was found that the shrinkage rate of concrete increases with increasing ambient temperature. When the temperature was high, the first 7 days chemical reaction rate was faster, and it lead to higher shrinkage rates. The early shrinkage would be expected under high temperature and low humidity environment. Chemical reactions and hydration rates reduced with age and so as

Table 5. Shrinkage rate at 30% humidity

Grade	Age			
	7 days	14 days	28 days	60 days
C30	430	610	1040	1130
C40	406	545	970	1052
C50	378	523	916	976

Units 10^{-6}

Table 6. Shrinkage rate at 60% humidity

Grade	Age			
	7 days	14 days	28 days	60 days
C30	127	240	512	564
C40	162	276	524	603
C50	180	302	550	638

Units 10^{-6}

Table 7. Shrinkage rate at 95% humidity

Grade	Age			
	7 days	14 days	28 days	60 days
C30	78	94	120	133
C40	124	138	160	175
C50	138	155	178	193

Units 10^{-6}

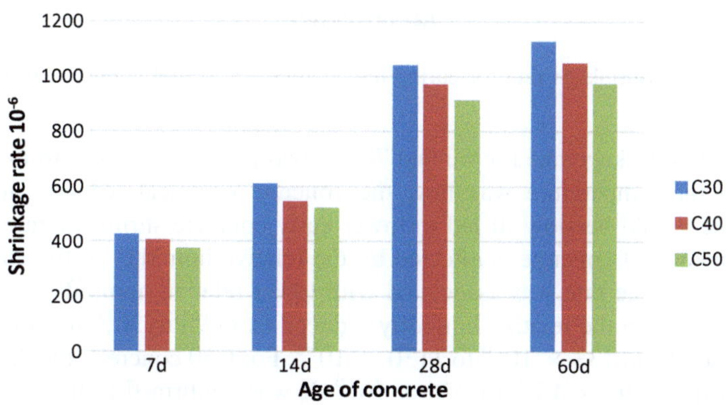

Fig. 4. Shrinkage rate for curing temperature 25 °C and humidity 30%

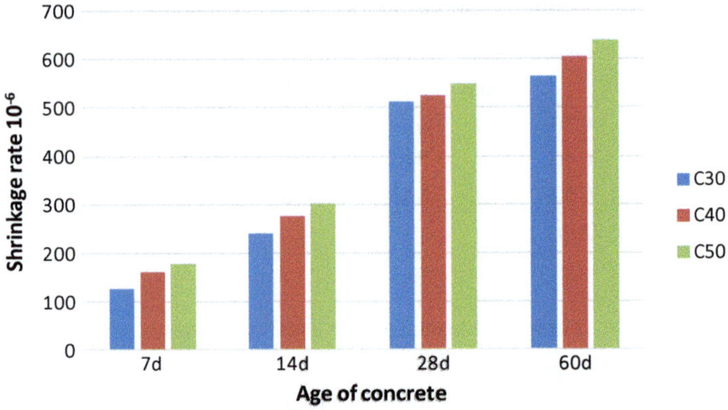

Fig. 5. Shrinkage rate for curing temperature 25 °C and humidity 60%

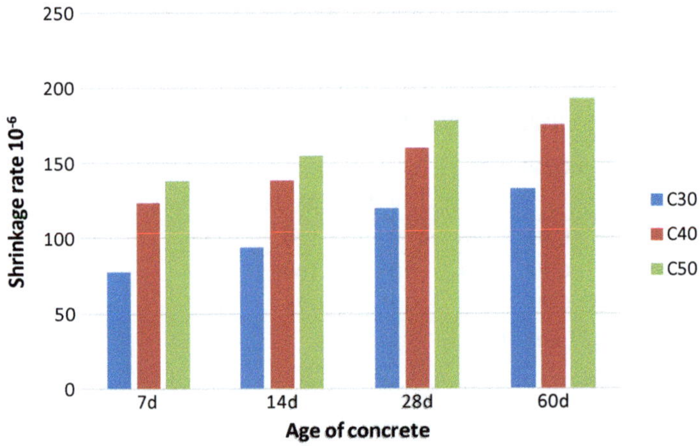

Fig. 6. Shrinkage rate for curing temperature 25 °C and humidity 95%

shrinkage rate. Shrinkage rate for the first 7 days was higher than those from 7th to 60th days. When the temperature was low, the concrete chemical reaction and the wet exchange rate would be slow. It led to lower early concrete shrinkage rate [5].

The shrinkage of concrete is affected by the relative humidity of the environment. The higher shrinkage rate was associated with lower relative humidity. For C30 concrete at 25 °C, when the relative humidity reduced from 95 to 30%, the dry shrinkage value increased from 133×10^{-6} to 1130×10^{-6}. For C50 concrete, the dry shrinkage value raised from 193×10^{-6} to 976×10^{-6}. It was confirmed in this study that the relationship between dry shrinkage and environmental relative humidity cannot be described by a simple linear relation [7].

When humidity was less than 60%, and the water dissipation rate of concrete was higher. In early life of concrete, when there was insufficient humidity, it would cause the water in the concrete to evaporate massively. It would impact hydration of cement due to loss of water. The dry shrinkage led to development of tensile stress in low strength state and caused cracks on the concrete surface. The durability index such as strength and impermeability of concrete would be adversely affected. Therefore, concrete curing immediately after placement was very important to ensure sufficient moisture during hydration as it would ultimately affect the performance of concrete [7].

6 Conclusions

Drying shrinkage happens mostly because of the reduction of capillary water by evaporation and the water in the cement paste. Effects of humidity, temperature, and cement grade on dry shrinkage were investigated in this study. It was found that humidity has the most profound effects on concrete dry shrinkage. To minimize shrinkage crack and improve concrete durability, the curing immediately after placement for the first 7 days was very important to ensure sufficient moisture during cement hydration. It is advisable to keep humidity higher than 60% such that the shrinkage of concrete can be minimized.

References

1. Xiao, R., Zhang, X., Le, J.: The study on the effect of gelled materials on shrinkage of concrete: concrete and cement products (2002)
2. Bai, K., Yu, H., Hao, J., Cao, W.: Mathematical model and drying shrinkage of high volume mineral admixture. J. Hangzhong Univ. Sci. Technol. (2008)
3. Leng, F., Zhou, Y., Wang, J.: Durability of Concrete and Its Test Evaluation Method. China Building Materials Industry Press, Beijing (2012)
4. Tan, X., Chen, W., Wang, L.: Experimental study on shrinkage properties of concrete at home and abroad. Shanxi Construct. **1**, 157–158 (2006)
5. Qian, X., Zhan, S., Zhou, F., Zhu, Y.: The effect of early curing time on the early shrinkage of concrete. J. Shenyang Univ. Architect. (Nat. Sci. Ed.) **7**, 610–613 (2007)
6. Tan, W.: Shrinkage and cracking of concrete and its evaluation and prevention. Tsinghua University
7. Wei Cao, R., Liu, B.: Effect of temperature and humidity concrete shrinkage test results. J. Ludong Univ. (Nat. Sci. Ed.) **27**(4), 376–379 (2011)
8. GB/T50082-2009, general concrete long-term performance and durability test method standard

Representing Collected Road Condition Data with Chernoff Faces for Evaluation of Pavement Conditions

Aioub H. Guha[1,2(✉)] and Gabriel J. Assaf[1]

[1] Department of Civil Engineering & Construction, École de Technologie Supérieure (ÉTS), University of Québec, 1100 Rue Notre-Dame Ouest, Montréal, QC H3C 1K3, Canada
Aioub.Guhai.1@ens.etsmtl.ca, Gabriel.Assaf@etsmtl.ca
[2] Department of Construction, Technical Faculty of Structural Engineering, Meslata, Libya
info@tefm.org

Abstract. Maintenance engineers have a problem because of the great quantity of data they must consider when evaluating kilometres of roads and because road roughness data measured with IRI (or possibly PCI) cannot be considered in isolation from other data such as Construction and Workmanship; Mix and Structural Design; Traffic; Climate (during and after construction). Incorporating these five datapoints into a single conventional graph would require a graph in five dimensions and such a graph would lose much of the benefits that a graph is meant to give (i.e. speed of data comprehension and readability). The Chernoff Face Method (CFM) is an iconic visualization technique used to code multi-variate data to simplify two-dimensional line drawings of faces so as to create a representation of the data that is more intuitively comprehensible to humans. The theory behind CFM is that humans have a special sensitivity to details in a facial representation that they do not have with data representations such as conventional graphic displays (e.g. line and bar graphs). The objective is to find an optimum balance between the volume of the data represented and the comprehensibility of the data. In particular, humans can discern eye size and eyebrows slant in faces with finer distinctions than they would be able to intuitively understand when reading another graph. This promises to be a means of displaying fine grained data for analysis by mapping data points to possible display variables such as: eye size, face shape, eyebrows slant angle (inward, outward, neutral). In order to map the IRI, construction, design, traffic and climate data to the CFM, a Multi-Criteria Analysis (MCA) model was used to determine the optimal number for the scale for each data category. The total number of Chernoff Faces for this problem was found to be 243, so that all five categories of data could be represented. Using CFM in conjunction with this data allows better evaluation of pavement distress.

Keywords: Chernoff face · International roughness index · Traffic level Climate zone · Structural design · Material & construction · Multi-criteria analysis

K. Zhang et al. (eds.), *Testing and Characterization of Asphalt Materials and Pavement Structures*, Sustainable Civil Infrastructures,
https://doi.org/10.1007/978-3-319-95789-0_13

1 Introduction

Planners and maintenance engineers are often confronted by the problem of how to understand and analyze large quantities of data. In the domain of roadwork, the data is relatively simple, pertaining to qualities such as road surface roughness, traffic levels, climatic zones, structural design and materials and construction. While these five basic qualities are relatively uncomplicated, if they are defined, for example, every 200 ms over many kilometres of roads, the data describing these small variations become overwhelming. The problem is worsened if, for each of those data-points, the engineer must either assign or read a number that represents the actual real world condition. Keeping this all in active human memory is an almost impossible task.

1.1 Benefits and Limits of Measuring Factors in Pavement Distress

Among the various things affecting pavement quality (and therefore ride quality) pavement distress is considered to be the most important (REF). This distress plays a cumulative role in long-term and incremental degradation caused by traffic and the environment, but this distress can be accurately measured with the International Roughness Index (IRI) (Von Quintus et al. 2001). Presently, to simplify this data, it is often indicated in terms of the short, word descriptions of the IRI where road surfaces are graded and referred to in general terms such as "smooth", "rough", and "very rough" (as in Table 1, row 1) in order to help planners remember the basic quality of a road. Pavement evaluation methods, such as the IRI and the Pavement Condition Index, PCI, the standard alternative to IRI, are already very sophisticated in terms of data collection but the data these methods collect are hard to communicate to engineers and planners. This project uses IRI because it is one the most common ranking systems for pavement surface smoothness and therefore ride quality (Shafizadeh and Mannering 2003). Note that the present proposal is not for a new means of measuring pavement distress but simply a method to communicate pavement condition findings that have already been recorded with IRI (or possibly PCI) alongside contributing factors influencing the results collected by the IRI to engineers in a more intuitive way.

Beyond this consideration, the IRI (or PCI) data on their own is not entirely sufficient to communicate everything that an engineer needs in planning maintenance; it would be much more useful to have the IRI data displayed alongside corresponding contributory data (e.g. climate data). This is because the distress from traffic and the environmental factors depends greatly on several factors. Following Adlinge and Gupta (2013) the factors under consideration in the present research are: Pavement Roughness (IRI); Construction and Workmanship; Mix and Structural Design; Traffic; Climate (during and after construction). A better understanding of how these factors interact would help predict the change in the road rating and the changes in the performance of the pavement (Haider and Chatti 2009). For example, with empirical data collected in Virginia, New Jersey, and Maryland, Zhou and Wang (2009) created a predictive model for IRI. The results of this model were then compared with similar models helping them to identify qualities that indicated pavement distress. These were evaluated in terms of standard criteria for the progression of roughness in pavement over time. Not all pavement distresses have an equal effect on future performance and they

Table 1. The levels of quality for each pavement condition factor (Visintine et al. 2015)

	Factors types	Status	Description
1	Pavement roughness IRI	Good (= "smooth")	IRI < 8 m/km
		Fair (= "rough")	$8 \leq$ IRI ≤ 4 m/km
		Poor (= "very rough")	IRI > 4 m/km
2	Materials and quality	Good	Material meets specifications
		Fair	Material does not always meet specifications
		Poor	Material often does not meet specifications
3	Construction and workmanship	Good	All work meets specifications
		Fair	Work does not always meet specifications
		Poor	Most work does not meet specifications
4	Mix and structural design and specifications	Match design	Meets specifications
		Fair	Mix and/or structural design recommendations does not met structural considered
		Does not met	No mix and/or structural design recommendations
5	Traffic level	Low volume	Fewer than 5000 vehicles per lane per day
		Medium volume	From 5000 to 20,000 vehicles per lane per day
		High volume	Over 20,000 vehicles per lane per day
6	Climate zone problems	None	No weather problems during or right after construction
		Few	Some weather problems during or right after construction
		Yes	Significant weather problems during or right after construction

therefore affected the reliability of these predictive models (Wang et al. 2007). Smith and Tighe (2004) used the Long Term Pavement Performance (LTPP) index to research what affected the development of pavement roughness and reported that regional environmental factors and asphalt design had the greatest influence on pavement quality.

Several research papers have summarized factors contributing to ride quality but they are limited because there are many obstacles to empirical data representation (Haider and Chatti 2009). This seemingly simple detail is important in the study of pavement deterioration, particularly when trying to correlate such things as environmental effects with pavement quality changes over time. As such, systems like IRI and

PCI are not, on their own, sufficient to communicate all the details that may be important in judging road quality, and planning maintenance.

1.2 Problems with Graphs and Data Representation

In order to detect patterns, clusters of effects and even outlying data points in a dataset, researchers have often used graphical representations. These help people see trends that are not obvious without such graphical representations or "graphs". In such graphs, a balance must be found between data volume and practical means of displaying the data. Nonetheless, indicating more details, in terms of numerical values, as conveyed in a standard graph (bar, line graph, etc.), is hard to remember and is generally considered to be unworkable. Furthermore, a standard graph starts to become hard to read beyond two dimensions (x, y), starting with a graph including three dimensions (x, y, z) and it becomes time consuming to read when there are more dimensions than this, losing many of the advantages of having a graph in the first place. Nonetheless, when engineers are examining many kilometres of roads and must make decisions as to the priority of what to repair sooner and what to repair later, they need more than just two data points; for example, understanding the structural design of a segment of the road can help in those repair decisions. Again, this problem is multiplied by the amount of data represented, i.e. kilometres of roads described every 200 m.

1.3 The Chernoff Face Method (CFM)

One innovative idea was published by Herman Chernoff, a statistician, as "The Use of Faces to Represent to Points in K-Dimensional Space Graphically," in 1973 in the Journal of the American Statistical Association (Chernoff 1973). His idea was to exploit the human brain's greater sensitivity to human faces in order to represent complex multivariate data and to assimilate it with other such data at the same time. As described in his 1973 paper, he encoded such multivariate data in the features of simplified human faces; the faces generated by this technique are now known as Chernoff faces and here as the Chernoff Face Method (CFM). In Fig. 2, several of these variables are assigned to a facial trait; from this, a face is generated for each case. In practical terms, Chernoff faces are two-dimensional images, often line drawings, each containing very specific features. Each detail of these facial features can be correlated to features in a dataset. Chernoff faces have been applied to visualization problems in areas such as intelligence analysis, information retrieval and data mining (Chernoff 1973) and using Chernoff faces may allow engineers to represent large datasets in a way that is intuitively accessible for people.

In order to take advantage of the human brain's natural capacities for recognition of various facial features, the options in the Chernoff faces have to be limited to stay within a range of realistic or semi-realistic expressions or characteristics. Therefore, the quantity of features that are adjusted is restricted to that which would make any face contrast with another. Nonetheless, some characteristics are more noticeable to us than others. The human capacity to notice small changes in these features can be ordered as follows: face size; face shape; nose length; mouth location; degree of curves of the smile, location; eye characteristics such as angles, shape, separation, width; location of pupil; width, angle, location of eyebrows. By means of these variables, an engineer can display and store many of these data in a single facial representation, i.e. as opposed displaying a data point as a point on a line graph, the engineer displays a Chernoff face with, for example, five data points within it. Nonetheless, this data could be greater than five and Kabulov and Tashpulatova (2010) found that, with enhancements such as hair, beards, and colour, they were able to display up to twenty dimensions in a single face. Therefore, CFM promises to be a means of displaying fine grained data for analysis by collecting data points such as: Pavement Roughness (IRI Materials and Quality); Construction and Workmanship; Mix and Structural Design; Traffic; Climate (during and after construction) (Adlinge and Gupta 2013). These will be mapped to display variables such as shape (everything), colour (face, eyes), size (all except face), spacing (eyes), slant angle (eyebrows, e.g. inward, outward, neutral). See Fig. 1. Mapping data to CFM promises to assist engineers in their recognition and recall of prominent features and patterns of data; these simple but powerful visualization tools can help projects to better understand variations in road pavement quality along the course of a road (Fig. 2).

Fig. 1. Examples Chernoff faces

As any conventional graph, using CFM helps recognize clusters and single instances of changes in the quality roads as well as the variables that affect such changes. For the simulations, five facial features were used to characterize the greatest variations in the factors proposed by Adlinge and Gupta (2013) (i.e. traffic loads, environment, pavement thickness, material). These were represented by facial characteristics including: nose width and length; mouth curvature; eyebrow length and angle. In order to speed processing, the facial variables that were not needed for the study were set as default values; these included width and shape of eyes and face.

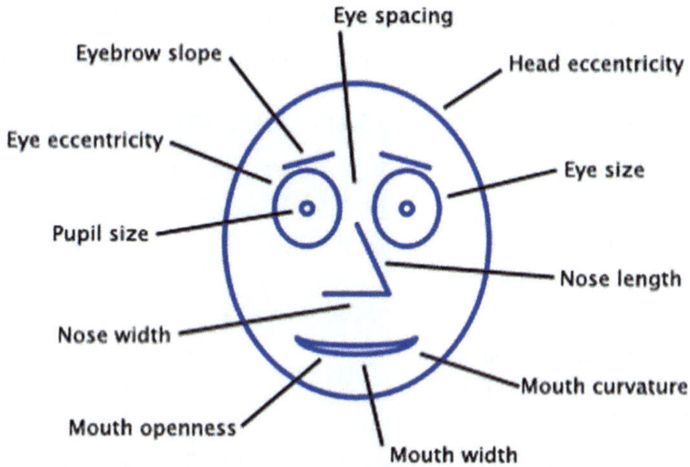

Fig. 2. Facial variations

1.4 Incorporating CFM with Multivariate Data

At this point the benefits with of using CFM in conjunction with IRI seem apparent. If the project only considers incorporating the IRI into the CFM, it could simply improve on the standard IRI designations of smooth, rough, and very rough. Initially this was the approach and, for the purposes of integrating the IRI into a broader scale, this study did the following equivalences: Good (= "smooth") IRI < 8 m/km; Fair (= "rough") $8 \leq$ IRI ≤ 4 m/km; Poor (= "very rough") IRI > 4 m/km. In order to evaluate the overall quality of the pavement, this study determined a scale based on common representations of pavement conditions. For practical purposes, this study initially chose a ten-point scale, following the scale discussed in the PASER Asphalt Roads Manual (Walker et al. 2002) where 10 is excellent and 1 is failed; see Table 2.

Each step of the PASER scale, as shown in Table 2 (adopted from the PASER manual), gives detailed criteria relating to the road's condition and whether or not it needs repair. 10 indicates a new construction; 9 is a "like new recent overly; 8 is a recent seal coat or new cold mix that requires little or no maintenance; 7 shows first signs of aging; 6 shows signs of aging and the road life could be extended with the seal coat; 5 indicates surface aging but with sound structural conditions; 4 has significant aging and would benefit from 2 inches or more of structural overlay; 3 needs patching and repair before a major overlay; 2 indicates severe deterioration and a need for reconstruction with extensive base repair; 1 indicates a failed pavement with a need for total reconstruction (from the PASER evaluation in Table 2). As such, the PASER scale in Table 2 is a practical needs-based evaluation scale. Details of the points proposed by Adlinge and Gupta (2013) follow.

1.4.1 Structural Design

Al-Suleiman and Shiyab (2003) used roughness data to build a model that estimated the length of the service life of existing roads in Dubai. They created two regression

Table 2. Pavement rating condition [based on the PASER scale (Walker et al. 2002)]

Surface rating	Visible distress	Photo surface distress	General condition
10 Excellent	None		New construction
9 Excellent	None		Recent overlay. Like new
8 Very good	All cracks sealed or tight (open less than ¼″)		Recent sealcoat or new cold mix. Little or no maintenance required
7 Good	Very slight or no raveling, surface shows some traffic wear		First signs of aging. Maintain with routine crack filling
6 Good	Sight to moderate flushing or polishing. Occasional patching in good condition		Shows signs of aging. Sound structural condition. Could extend life with sealcoat
5 Fair	Moderate to severe raveling (loss of fine and coarse aggregate). Extensive to severe flushing or polishing. Some patching or edge wedging in good condition		Surface aging. Sound structural condition. Needs sealcoat or thin non-structural overlay (less than 2″)
4 Fair	Severe surface raveling. Multiple longitudinal and transverse cracking with slight raveling. Longitudinal cracking in wheel path		Significant aging and first signs of need for strengthening. Would benefit from a structural overlay (2″ or more)
3 Poor	Closely spaced longitudinal and transverse cracks often showing raveling and crack erosion. Patches in fair to poor condition. Moderate rutting or distortion (1″ or 2″ deep)		Needs patching and repair prior to major overlay. Milling and removal of deterioration extends the life of overlay

(*continued*)

Table 2. *(continued)*

Surface rating	Visible distress	Photo surface distress	General condition
2 Very poor	Alligator cracking (over 25% of surface). Severe distortions (over 2″ deep). Extensive patching in poor condition. Potholes	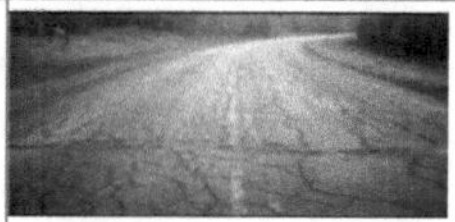	Severe deterioration. Needs reconstruction with extensive base repair. Pulverization of old pavement is effective
1 Failed	Severe distress with extensive loss of surface integrity		Failed. Needs total reconstruction

simulations that proved to be statistically significant and that successfully estimated how much service life remained in a slower high traffic lane and an express traffic lane. Hall et al. (2003) studied how control sections compared with test sections of the road with regards to long-term roughness, cracking and rutting. With Canadian LTPP data Raymond et al. (2002) analyzed the progress of pavement roughness in the first eight years of the lifespan of a new road construction. Their findings indicated that overlay thickness, location, and preexisting cracking all played sizeable roles in the progression of pavement roughness. Prior to resurfacing, factors such as overlay thickness, pavement roughness and how much the surface had been prepared all played a statistically significant effect (95% significance) on the degree of roughness of a newly resurfaced pavement (Raymond et al. 2003). Perera and Kohn (2006) replicated an earlier study tracking roughness on resurfaced pavements showing that the IRI of the pavements depended on the IRI prior to overlay of a new layer, as well as the thickness of the new overlay. Their conclusions did not find that type of asphalt pavement or milling before a new overlay was important to the later progression of pavement roughness, i.e. while a thicker overlay resulted in a surface with a lower IRI (smoother) thickness did not affect IRIs for sections laid over unbound bases (Rahim et al. 2009).

Within the realm of structural design, moisture is the single greatest cause of deterioration of an asphalt road, at all levels of construction. Which is to say, it degrades the surface asphalt level but also the subgrade. Water penetrates the pavement in three general ways: by way of holes and cracks in the top layers; laterally, through the subgrade itself; with capillary action from the water table below the road. Moisture causes particles to be lubricated, which itself causes the interlocking between the particles to deteriorate and for particle displacement. This causes pavement failure (Adlinge and Gupta 2013).

Whether or not moisture has entered the subgrade, ultimately, it is these soils of the subgrade, under the pavement, that support the axle loads. When the subgrade is not sufficient to support these axle loads, the road again becomes too prone to movement and this greater movement eventually results in pavement failure. In cases where there are variations in the subgrade composition that are not or cannot be sufficiently compacted or otherwise compensated for during the pavement design or construction, it can be expected that the pavement will perform poorly or for a shorter time (Adlinge and Gupta 2013).

1.4.2 Climate Zone

Haider and Chatti (2009) found that another important factor affecting IRI was environmental factors. In climates that did not freeze, the greater the variation and degree of moisture in the subgrade, the greater and the faster the increase in pavement roughness. In freezing sites, they found a parallel, where freezing and moisture combined to influence the progression of roughness indices. As might be expected, sites that experienced high degrees of both moisture and freezing indicated the highest reported roughness levels (Haider and Chatti 2009). Controlling for subgrade types, thinner overlays in areas with high levels of moisture and freezing had higher roughness progression than thicker ones (Tighe 2002). Nonetheless, roughness progression was the highest in sites with thin overlays, fine subgrade, situated in low-freeze but wet areas (Tighe et al. 2000). Although the condition of a pavement before overlay has an important effect on the as-built roughness of a road (Raymond et al. 2003), there is, as yet, no study of how overlays on concrete or asphalt roads progress in terms of roughness.

1.4.3 Traffic Levels

The greatest influence on pavement lifespan and performance is traffic. Specifically, the effects of traffic that play the greatest role in the deterioration of pavement are factors such as the configuration, loading magnitude, and load repetitions by vehicles, particularly heavy trucks (Pais et al. 2013). Every axle load contributes to this deterioration and the standard single-axle load is defined as 80 KN (E80). This defines the deterioration of the road per axle per pass. As such, an asphalt pavement is planned for a given quantity of standard axle loads (i.e. E80s) before it falls below the service level for which it was designed (Adlinge and Gupta 2013).

1.4.4 Quality of Construction

Pavement is directly affected by material quality, moisture levels (during and after construction), degree of compaction, and correct final layer thickness. Due to the critical nature of these factors, skilled design, construction, and quality control inspection crews are very important for road construction. Table 1, shows the levels of quality for each factor for pavement (Visintine et al 2015).

2 Results

The initial project idea was that the PASER-inspired ten-point scale should be sufficient to determine the factors influencing pavement quality mentioned above. These data were then graded according to whether they are good, fair, or poor except in the case of Traffic, where they were graded as: low volume, medium volume, high volume. These data were then going to be mapped to the CFM to provide a better understanding of the existing pavement quality so engineers could make better predictions about what is happening below the pavement surface. An initial Table 1, was created to show the qualities from the IRI as they had been ranked with Traffic Level, Climate Zone, Structural Design, and Material & Construction on a ten-point scale derived from the PASER Manual scale, where 1 = Failed and 10 = Excellent.

Nonetheless, because the project has as an objective the representation of multivariate data, it soon became clear that the dataset would be much too large and therefore unworkable. If many of the variables (e.g. IRI, structural design) were assigned a ten-point scale, the output would give us an outcome of 105 faces. This is clearly not a satisfactory result.

2.1 Multi-criteria Analysis Model

With the project objective of displaying both the IRI data and the data representing these last four areas (structural design, climate zone, traffic levels and quality of construction) a reduced scale for each of these had to be determined. These new scales were chosen based on output from a Multi-Criteria Analysis model (MCA). MCA is a way to make an informed choice when there are many small data points that might otherwise be too complex to navigate and hard to get an actionable outcome. Therefore the MCA is set up so that many small choices are made between two variables and this is repeated until a final outcome is determined.

Using MCA, the final scale is comprised of 243 faces. Following the logic that a viewer can easily discern between three states of a feature, there are only three states to each feature. For example, with the mouth, a poor road surface is mapped to a frowning mouth; an adequate road surface is mapped to a neutral mouth (shown with a flat line); an excellent road surface is mapped to a smiling mouth.

The mappings, e.g. of IRI to Mouth, and Traffic Level to Nose, are somewhat arbitrary; because IRI is the determining scale that gives us an overall idea of the

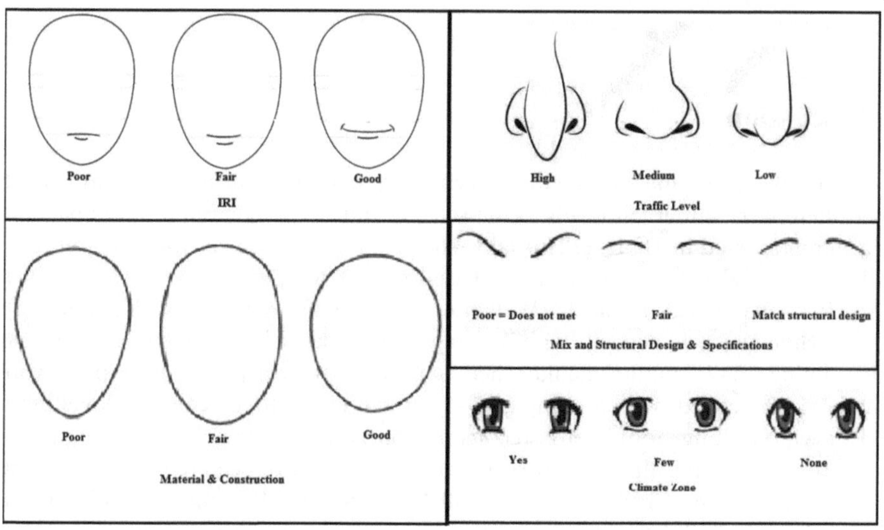

Fig. 3. Example of shapes to each feature according to the factors in Table 3

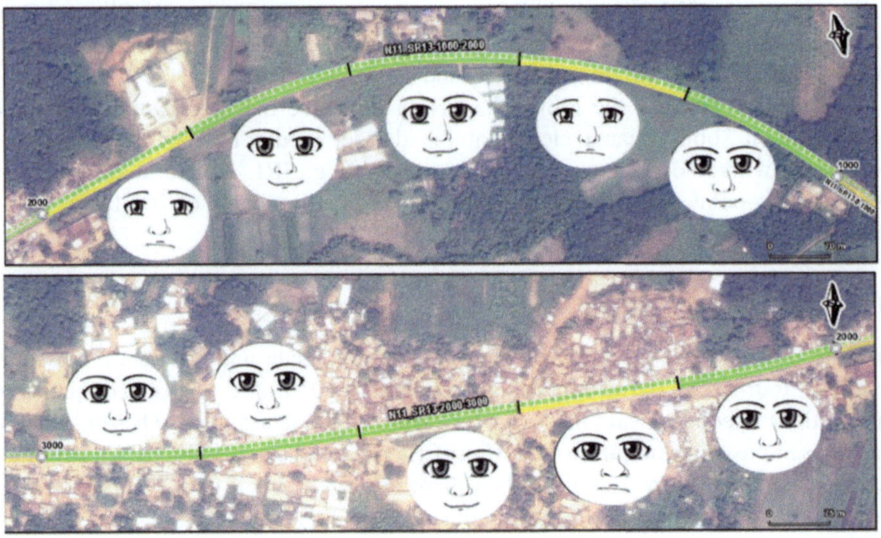

Fig. 4. Examples of Chernoff faces showing the factors affecting the surface pavement condition

quality of the road surface, it is mapped to the mouth, on the assumption that a mouth gives the most obvious impression of an emotional state, i.e. it would make little intuitive sense to have a bad road surface and a happy expression on the mouth, as it is shows in Fig. 3.

Figure 4 shows a GIS-generated map with the Chernoff Faces superimposed on it, that indicate the road quality alongside the other data points from Table 3. The image in Figs. 3 and 4 is an example of the standard output of the CFM and could be easily distributed to planners and maintenance engineers making decisions as to what repairs to schedule.

3 Discussion

Evaluating the condition of pavement surfaces is complicated in many ways. In this paper, there were initial ten levels in the scale to reflect the condition of the pavement surface, represented in Table 2. This table shows the initial choices for this project with all the levels and the pavement rating based on the ten-point scale discussed in the "PASER Asphalt Roads Manual" (Walker et al. 2002). For example, with a rating of 9–10, no maintenance is needed; with a rating of 8, little or no maintenance is needed; with a ratings 6–7, routine maintenance is needed; with a rating of 5, preservative treatments are needed; with a rating of 3–4, structural improvements and levelling are needed; with a rating from 1 to 2, reconstruction is needed.

These two systems, the PASER scale and the CFM, have been represented in a new scale, called here the CFPSC, the Chernoff Faces Surface Pavement Condition scale, illustrated in Table 3. The five categories, seen as column designations, represent the data variables with Chernoff faces. These can then be correlated to the pavement rating

Table 3. Pavement rating condition ten-point scale

Segment		IRI m/km		Traffic level		Climate zone		Structural design		Material and construction	
		Mouth		Nose		Eye		Eyebrow		Face	
Good	5	10	5	10	5	10	5	10	5	10	5
		9		9		9		9		9	
		8		8		8		8		8	
Fair	3	7	3	7	3	7	3	7	3	7	3
		6		6		6		6		6	
		5		5		5		5		5	
Poor	1	4	1	4	1	4	1	4	1	4	1
		3		3		3		3		3	
		2		2		2		2		2	
		1		1		1		1		1	

conditions, seen as rows, and then correlated with the found levels of pavement deterioration.

As explained in the Results, if the PASER scale were directly implemented into a Chernoff Face Method, there would be as many as 10^5 face icons. These are clearly too many options for the practical effects that are needed for a successful implementation of the Chernoff Face Method. Therefore, the 10-point scale seen in Table 2 was reduced to three levels for each of the five categories as shown in Table 3, using a Multi-Criteria Analysis model (MCA).

This new scale takes the numerical and the word values from Table 2 and gives new designations in Table 3; these new designations are much simpler but still capture the information required. The Table 2 values of 10, 9, 8 indicate that the pavement condition is "excellent" or "very good"; these are simplified in Table 3 and coded as "green/5/good" (where the number corresponds to the value needed for the MCA). The same equivalences are made from Table 2 where grades 7, 6, 5 are indicated in Table 3 as "yellow/3/fair". In the final equivalence from Tables 2 to 3, Table 2 grades 4, 3, 2, 1 is indicated in Table 3 as "red/1/poor". (This conversion process relates to the present project only and would not be entered into for the end user, because it is unnecessary and possibly confusing).

The final Table 3 ratings are determined according to the need for maintenance or repair. Where roads coded "green/5/good" are in good condition and need little to no maintenance. Roads coded "yellow/3/fair" show some surface aging and may benefit from some non-structural maintenance. Roads coded "red/1/poor" require more work, ranging from structural overlay to total reconstruction. These, in turn, indicate the ride quality of the pavement surface.

Figure 3 clearly shows the features of Chernoff faces related to the pavement surface. As indicated above, the variations in faces have been determined, in part, using a Multi-Criteria Analysis model (MCA), using excel. The MCA determined that 243 faces are sufficient to show all the conditions discussed in Table 3. Figure 4 was generated with the Geographic Information System software (GIS), although other software could also be used. To prepare to output an image, such as in Fig. 4, the 243 faces need to be input into the GIS; in Fig. 4, the road was divided into 200 m segments; according to the condition of the road, the GIS maps the correct Chernoff face to each stretch of road.

4 Summary and Conclusions and Further Research

Pavement rating is an important part of an agency's approach to evaluating road performance. This paper determines how the Chernoff faces help engineers to evaluate the performance of pavement according to many factors, such as pavement condition, construction quality, quality of materials, traffic, and climate. The results clearly show that the deviations from good conditions can have a dramatic effect on the acceptable service life of the pavement, where pavement roughness as determined by the IRI. This new technique will avoid the problem of conflating general pavement quality of a large area of pavement with the quality of a small area contained within the greater pavement surface area.

The challenge for the project now is to determine the best output values for the various shapes and details of the faces so that they represent data points that can be easily read in the field. For example, if the CFPSC (The Chernoff Faces Surface Pavement Condition Scale) relies too much on colour, it may prove difficult to read with a black and white printout; likewise, if it relies too much on size, the data may prove to be harder to judge quickly on a low quality computer screen or printout. Therefore, these variations are being tested and considered with care.

It is clear that the Chernoff Faces Surface Pavement Condition scale (CFPSC), once fine tuned, is superior to crude descriptions such as good, fair, or poor condition.

References

Al-Suleiman, T.I., Shiyab, A.M.: Prediction of pavement remaining service life using roughness data—case study in Dubai. Int. J. Pavement Eng. **4**(2), 121–129 (2003)

Aultman, L., Jackson, E.: Model relating pavement quality measures. Journal of the Transportation Research Board, Washington (2004)

Bae, A., Stoffels, S.M., Antle, C.E., Lee, S.W.: Observed evidence of subgrade moisture influence on pavement longitudinal profile. Can. J. Civ. Eng. **35**(10), 1050–1063 (2008)

Byrum, C.R., Perera, R.W.: The effect of faulting on IRI values for jointed concrete pavements. In: Eighth International Conference on Concrete Pavements (2005)

Chernoff, H.: The use of faces to represent points in k-dimensional space graphically. J. Am. Stat. Assoc. **68**(342), 361–368 (1973)

Haider, S.W., Chatti, K.: Effects of design and site factors on roughness development in flexible pavements. J. Transp. Eng. **135**(3), 112–120 (2009)

Hall, K., Correa, C., Simpson, A.: Performance of rigid pavement rehabilitation treatments in the long-term pavement performance SPS-6 experiment. Transp. Res. Rec. J. Transp. Res. Board **1823**, 64–72 (2003)

Ismail, M.S.N., Ghani, A.N.A. An overview of road damages due to flooding: case study in Kedah state, Malaysia. In: AIP Conference Proceedings, vol 1892, no. 1, p. 190001). AIP Publishing (2017)

Pais, J.C., Amorim, S.I., Minhoto, M.J.: Impact of traffic overload on road pavement performance. J. Transp. Eng. **139**(9), 873–879 (2013)

Perera, R.W., Kohn, S.D.: Ride quality performance of asphalt concrete pavements subjected to different rehabilitation strategies. In: Airfield and Highway Pavement: Meeting Today's Challenges with Emerging Technologies, pp. 789–800 (2006)

Rahim, A.M., Fiegel, G., Ghuzlan, K., Khumann, D.: Evaluation of international roughness index for asphalt overlays placed over cracked and seated concrete pavements. Int. J. Pavement Eng. **10**(3), 201–207 (2009)

Raymond, C.M., Haas, R.C., Rothenburg, L.: Development of a C-LTPP pavement performance model for asphalt overlays. In: Proceedings of the Annual Conference—Canadian Society for Civil Engineering, Canadian Society for Civil Engineering, pp. 2553–2562 (2002)

Raymond, C.M., Tighe, S.L., Haas, R., Rothenburg, L.: Analysis of influences on as-built pavement roughness in asphalt overlays. Int. J. Pavement Eng. **4**(4), 181–192 (2003)

Shafizadeh, K., Mannering, F.: Acceptability of pavement roughness on urban highways by driving public. Transp. Res. Rec. J. Transp. Res. Board **1860**, 187–193 (2003)

Smith, J., Tighe, S.: Assessment of overlay roughness in long-term pavement performance test sites: Canadian case study. Transp. Res. Rec. J. Transp. Res. Board **1869**, 126–135 (2004)

Tighe, S., Haas, R., Ningyuan, L.: Roughness progression of asphalt overlays in the C-SHRP's LTPP study. In: Proceedings of the 2000 CSCE Annual Conference, London, Ont, pp. 7–10 (2000)

Visintine, B.A., Hicks, R.G., Cheng, D., Elkins, G.E.: Factors affecting the performance of pavement preservation treatments. In: 9th International Conference on Managing Pavement Assets (2015)

Von Quintus, H., Eltahan, A., Yau, A.: Smoothness models for hot-mix asphalt-surfaced pavements: developed from long-term pavement performance program data. Transp. Res. Rec. J. Transp. Res. Board **1764**, 139–156 (2001)

Walker, D., Kummer, S., Entine, L.: Pavement Surface Evaluation and Rating; PASER Asphalt Roads Manual. Transportation Information Center, University of Wisconsin-Madison (2002)

Wang, K., Li, Q., Hall, K., Elliott, R.: Experimentation with gray theory for pavement smoothness prediction. Transp. Res. Rec. J. Transp. Res. Board **1990**, 3–13 (2007)

Zhou, G., Wang, L.: . Analysis of flexible pavement distresses on IRI model. In: Pavements and Materials: Modeling, Testing, and Performance, pp. 150–160 (2009)

Reflective Cracking in Composite Pavements—A Case Study

Rongyao Du(✉)

Guangxi Communications Investment Technology Co. Ltd., Nanning City 530000, Guangxi Province, People's Republic of China
191604978@qq.com

Abstract. In this study, ten sections of different composite pavement structures in southern China were tested. Study in fields included distress survey, characteristic of reflective cracks with drilled core, and Falling Weight Deflectometer (FWD) testing. The Top-down crack was found, and the thick overlay and SMA pavement was found to be favorable to control the reflective cracks.

1 Introduction

Composite pavement, commonly consisting of asphalt concrete (AC) overlay and Portland cement concrete (PCC), is the main type of pavement used in expressway in southern China. Reflective cracking is one of the major distresses in AC overlays on jointed cement concrete pavements (JCCP), which is without rebar dowel in all joints. The basic mechanism for reflective cracking in the overlay is strain concentration at the bottom and the vertical horizontal movements of the underlying PCC (Portland Cement Concrete) slab at the cracks/joints in the existing pavement (An et al. 2014). Li et al. (2011) analyzed the fatigue damage characteristics, cracking mechanism and crack propagation of asphalt overlay on JCCP under vehicle loading, using the theories of damage mechanics, fracture mechanics and simulation method; and indicating that the asphalt concrete overlay subjected to partial loading on the joints is prone to fatigue cracking under repeated load due to overlay being under tensile stress. Zeng et al. (2013) analyzed the tensile and shear stresses by modeling the AC overlay on continuously reinforced concrete pavement (CRCP) by using the finite element software, under symmetric and eccentric loading. Chen et al. (2007) analyzed the stress intensity factors of AC overlay on CRCP using the general finite element code ABAQUS.

The study demonstrates the field test results (Du et al. 2016), and analyses the mechanism of reflective cracking.

2 AC Overlays

The JCCP of ten tested sections located in the Expressway of Liuzhou to Nanning Express way were constructed in 1999, with the thickness of 26 cm and design flexural strength of 5 MPa, and the AC layers were overlaid between 2005 and 2010. The half-width pavement of the expressway was composed of two lanes, and the inner lane

K. Zhang et al. (eds.), *Testing and Characterization of Asphalt Materials and Pavement Structures*, Sustainable Civil Infrastructures,
https://doi.org/10.1007/978-3-319-95789-0_14

is constrained to light vehicles, while the outer lane for heavy trucks. The overlays are composed of AC-13, SMA-13, stress absorbing layer and geotextile. Table 1 shows the ovarlay structures of ten sections.

In August 2016 site test was conducted to evaluate the structural strength of ten composite pavements (Du et al. 2016). Site work in fields included pavement distress

Table 1. Overlay structure of the ten tested sections

Section number	Time in service (year)	Length (km)	Aspahlt overlay structures	Overlay thickness (mm)
1	11	1.5	4 cm AC-13 modified asphalt mixture + 5.5 cm natural asphalt AC-25 + 2 cm Asphalt sand stress absorbing layer	115
2		1.5	4 cm AC-13 modified asphalt mixture + 5 cm AC-25 modified asphalt mixture + 2.5 cm Strata stress absorbing layer	115
3		2	4 cm AC-13 SBS modified asphalt concrete +5.5 cm AC-25 SBS modified asphalt concrete + 1.5 cm Asphalt sand stress absorbing layer	110
4		1.5	3 cm SMA-13 modified asphalt mixture + 5 cm AC-25 modified asphalt mixture + Geotextile +1.5 cm Asphalt sand stress absorbing layer	95
5		1.5	4 cm SMA-13 modified asphalt mixture +6 cm AC-25 modified asphalt mixture + Geotextile +1.5 cm Asphalt sand stress absorbing layer	115
6	10	2	3.5 cm SMA-13 modified asphalt mixture +8.3 cm AC-25 modified asphalt mixture + Geotextile +1.5 cm Asphalt sand stress absorbing layer	131
7		5	3.5 cm SMA-13 lake-asphalt modified asphalt mixture +7 cm AC-25 lake asphalt modified asphalt mixture + Geotextile +1.5 cm Asphalt sand stress absorbing layer	120
8		5	3.5 cm SMA-13 SBS modified asphalt mixture +7 cm AC-25 SBS modified asphalt concrete + Geotextile +1.5 cm SBS modified asphalt sand stress absorbing layer	120

(*continued*)

Table 1. *(continued)*

Section number	Time in service (year)	Length (km)	Aspahlt overlay structures	Overlay thickness (mm)
9	5	2	4 cm SMA-13 SBS modified asphalt mixture +SBS modified asphalt tack-coat (2 kg/m^2) + 6 cm AC-20 high modulus asphalt concrete +3 cm AC-10 SBS modified asphalt concrete + Tack coat (SBS modified asphalt-2 kg/m^2)	130
10		12.4	4 cm AR-AC-13 + Tack coat (asphalt rubber-2.2 kg/m^2) + 6 cm AR-AC-20 + Tack coat (asphalt rubber -0.4 kg/m^2) +2 cm AR-AC-10 + Tack coat (asphalt rubber-0.4 kg/m^2)	120

survey, feature description of reflective crack features by drilling core bridging the visible reflective cracks, detecting the deflections of joints between two slabs using the Falling Weight Deflectometer, laboratory experiments were performed to test the bitumen content, voids content and aggregate gradation of AC layer using the drilled cylinder cores.

3 Pavement Condition

The technical report (Du et al. 2016) shows that typical distresses of overlays are reflective cracking, block cracking, rutting and pitting. The distress of the asphalt overleys surveyed in the year of 2007, 2008, 2009 (Du et al. 2016), respectively, are analysed and listed in Table 2. It is seen that the overlays of Sections-9 and 10 had no other distress except ruts. The density of distress is defined to characterize pavement performance, and correlated to deflection and overlay type on. Density of distress is calculated below (Table 3):

$$\begin{aligned}\text{The density of distress} = 100 \times (&\text{the area of the cracking } + \text{ area of patch} \\ &+ \text{ area of block cracking})/\text{the area of the pavement}\end{aligned}$$

Where, the area of the cracking = quantity × 3.75 m × 5 m (The 5 m is the influence width of the crack); the area of patching = the actual area of patching; the area of block cracking = 2× the actual area of patching (2 is the influence factor because the disease repair needs to milling one of the lanes of the pavement)

The AC overlays of sections 1, 2 and 3, whose pavement structures are composed of AC-13 SBS modified asphalt concrete upper layer, AC-25 natural asphalt concrete

Table 2. Distresses of eight sections of overlays

Site number	Distress types	Unit	Date of survey				The density of distress of 2016 (%)
			2007	2008	2009	2016	
1	Pit slot	Quantity	6	10	16	/	34.2
	Transverse cracking	Quantity	4	17	28	20	
	Patching	m^2	0	0	0	413	
	Block cracking	m^2/quantity	0	22.5/3	97/17	0	
2	Pit slot	Quantity	9	10	15	0	74.0
	Transverse cracking	Quantity	2	16	33	24	
	Patching	m^2	0	0	0	970	
	Block cracking	m^2/quantity	0	117/15	379/68	28/7	
3	Pit slot	Quantity	2	2	9	0	52.0
	Transverse cracking	Quantity	0	5	10	3	
	Patching	m	0	0	0	490	
	Longitudinal cracking	m	0	0	0	11	
	Block cracking	m^2/quantity	0	0.15/1	3.2/1	28/4	
4	Pit slot	Quantity	2	2	8	0	3.3
	Transverse cracking	Quantity	0	0	0	10	
	Patching	m	0	0	0	0	
	Longitudinal cracking	m	0	0	0	6	
	Block cracking	m^2/quantity	0	0	0	0	
5	Pit slot	Quantity	2	2	8	0	6.7
	Transverse cracking	Quantity	0	0	0	24	
6	Pit slot	Quantity	0	0	0	0	4.6
	Transverse cracking	Quantity	0	0	0	46	
	Patching	m	0	0	0	0	
	Longitudinal cracking	m	0	0	0	6	
	Block cracking	m^2/quantity	0	0	0	0	
7	Pit slot	Quantity	0	0	0	0	2.7
	Transverse cracking	Quantity	0	0	0	11	
8	Pit slot	Quantity	0	0	0	0	3.5
	Transverse cracking	Quantity	0	0	0	7	
	Longitudinal cracking	m	0	0	0	6	
	Block cracking	m^2/quantity	0	0	0	0	

or SBS modified asphalt concrete subsurface, and the bottom layer of stress absorbing layer, have been in service for 11 years, to the testing date. It is seen that distress densities of the three sections are higher than that of other sections. The main distresses surveyed in 2016 are transverse reflective cracks, and block cracks.

Table 3. The average thickness of overlay measured from the cores

Site number	Number of cores	Average thickness (mm)	Design thickness value (mm)
1	5	120	115
2	4	106	115
3	3	102	110
4	5	101	95
5	5	102	115
6	3	130	131
7	4	111	120
8	4	118	130
9	4	130	120
10	5	121	120

Sections 4, 5, 6, 7 and 8, whose overlay structures are composed of SMA-13 SBS modified asphalt mixture, AC-25 SBS modified asphalt concrete, geotextile and stress absorbing layer, with thickness of 9.5–13.1 cm, have been open to traffic for 10 years, to the testing date, suffering lower density of distress than sections 1, 2 and 3.

The field distress survey result indicated that the overlay with the upper layer of AC-13 SBS modified asphalt concrete suffered much larger amount of block cracking than the overlay with the upper layer of SMA-13 SBS modified asphalt mixture. It means that the upper layer of AC-13 had experienced fatigue failure, while the upper layer of SMA-13 had not.

4 Reflective Crack Feature

Core drilling was carried out to feature the reflective cracking. Some of the cores bridging the visible reflective cracks, while the other bridging the invisible joints identified from the slab side, were dilled down to the surface of concrete slab, and the thickness of cores were measured. The observation focused on identifying whether the reflective crack is Top-Down or not. The reflective crack is classified into two groups based on the scaling damage: low scaling crack and high scaling crack.

All the cores were tested in laboratory for further investigation. The laboratory work included the void content, bitumen content and aggregate gradation.

It is seen from Table 4 that all the reflective cracks were wide at the top and narrow at the bottom, and no breakage occurred in the geotextile and stress absorbing layer, indicating that the reflective cracks of Guangxi province whose whether is hot and humid, initiate from the top down to bottom (Figs. 5 and 6).

Fig. 1. Low-scaling transverse reflective crack

Fig. 2. The location of slab joint

Fig. 3. High-scaling transverse reflective crack

Fig. 4. The location of slab joint

Fig. 5. Low-scaling block crack

Fig. 6. The location of slab joint

5 Deflection Test Using FWD

Deflections of slab ends of joints and cracks were tested using FWD to evaluate the load transfer coefficient of the old PCC (Du et al. 2016). Only the outer lane was tested. Load transfer coefficient d_{je} is calculated using the equation below:

$$d_{je} = d_u / d_l \times 100$$

where, d_u is the deflection at unloaded slab end of the joint; while d_l is that at loaded slab end of the joint.

Deflections were randomly tested for the overlays with no visible reflective cracks. The representative deflection X' is evaluated by:

$$X' = \bar{X} + S\frac{t_a}{\sqrt{N}}$$

where, $\bar{X}$—the average of deflection of the section; S—standard deviation; $\frac{t_a}{\sqrt{N}}$—the coefficient of assurance.

Comparison with the sections 1, 2 and 3 (Table 4) show that at the same or similar pavement structure, there is a certain correlation between deflection and the density of distress of the section. From the section 3 (Table 5) and section 4 (Table 6).

Analysis on the Tables 4, 5, 6, 7, 8 and 9 indicates that the SMA overlay has much lower density of distress than the AC overlay with the similar deflection.

Table 4. Characteristics of the reflective cracks

Core location	Number of cores	Characteristics of the cracks
Low-scaling transverse reflective crack (Fig. 1)	7	Obvious cracks can be seen on the upper layer. The cracks were wide at the top and narrow at the bottom. The cores were corresponded to the slab joints (Fig. 2)
High-scaling transverse reflective crack (Fig. 3)	5	Obvious cracks can be seen on the upper layer and middle layer of the cores except the stress absorbing layer. The cracks were wide at the top and narrow at the bottom. The cores were corresponded to the slab joints (Fig. 4)
Low-scaling block crack	3	Obvious cracks can be seen on the upper layer of the cores. The cracks were wide at the top and narrow at the bottom. The cores were corresponded to the slab joint
Slab joint	5	No cracks were observed in the overlay in all cores
Low-scaling longitudinal reflective crack	1	Obvious cracks can be seen on the upper layer. The cracks were wide at the top and narrow at the bottom

Table 5. Distress density of 2016 and the deflection test results

Site number	The density of distress of 2016 (%)	Test lane and location	The representative deflections (0.01 mm)
1	34.2	Outer lane, tested at random location	11.3
2	74	Outer lane, tested at random location	14.2
3	52	Outer lane, tested in slab joint	11.8
		Outer lane, tested in the middle of slab	7.1
4	3.3	Outer lane, tested in slab joint	10.9
5	6.7	Outer lane, tested in slab joint/crack	18.5
6	4.6	Outer lane, tested in slab joint	9.2
		Outer lane, tested in the middle of slab	7.9
7	2.7	Outer lane, tested randomly	7.2
8	3.5	Outer lane, tested randomly	7.2
9	0	Outer lane, tested randomly	9

Table 6. Deflections and load transfer coefficients of joints/crack of section 3

Location	Lane tested	d_l (0.01 mm)	d_u (0.01 mm)	d_{je}
1	Outer lane	11.48	1.3	11.3
2	Outer lane	9.03	3.4	37.7
3	Outer lane	12.25	6.4	52.2
4	Outer lane	5.6	3.9	69.6
5	Outer lane	13.09	5.8	44.3
6	Outer lane	8.4	3.2	38.1
7	Outer lane	6.79	1.9	28.0
8	Outer lane	11.13	5.1	45.8
9	Outer lane	7.07	4.5	63.6
10	Outer lane	14.98	4.5	30.0

Table 7. Deflections and load transfer coefficients of joints/crack of section 4

Number	Lane tested	d_l (0.01 mm)	d_u (0.01 mm)	d_{je}
1	Outer lane	7.56	4.4	58.2
2	Outer lane	8.16	3.6	44.1
4	Outer lane	9.9	3.3	33.3
5	Outer lane	8.94	4.5	50.3
6	Outer lane	11.7	4.5	38.5
7	Outer lane	9.6	4.4	45.8
8	Outer lane	9.72	4.2	43.2
9	Outer lane	11.4	4.1	36.0
10	Outer lane	10.14	3.5	34.5
11	Outer lane	10.08	3.5	34.7
12	Outer lane	12	7.8	65.0
13	Outer lane	10.86	5.6	51.6
14	Outer lane	12.42	4.3	34.6

Table 8. Deflections and load transfer coefficients of joints/crack of section 5

Number	Lane tested	d_l (0.01 mm)	d_u (0.01 mm)	d_{je}
1	Outer lane	11.925	6.1	51.15
2	Outer lane	12.375	4.9	39.60
3	Outer lane	16.575	6.3	38.01
3	Outer lane	18.6	8.4	45.16
5	Outer lane	19.8	5.7	28.79
6	Outer lane	15.9	3.6	22.64

Table 9. Deflections and load transfer coefficients of joints of section 6 (no cracks)

Location	Lane tested	d_l (0.01 mm)	d_u (0.01 mm)	d_{je}
1	Outer lane	7.67	5.4	70
2	Outer lane	9.425	6.6	70
3	Outer lane	7.605	5.7	75
4	Outer lane	9.1	7.5	82
5	Outer lane	8.45	5	59
6	Outer lane	8.37	6.7	80
7	Outer lane	7.28	4.9	67
8	Outer lane	8.255	6.7	81
9	Outer lane	8.515	6.2	73
10	Outer lane	11.23	9.9	88

Table 9 shows that section 6 with low deflections and high joint load transfer coefficients had no cracks, whereas section 5 (Table 8) with high deflection and joint load transfer coefficient suffered from severe transverse cracking.

6 Mechanism of Reflective Cracking

As presented previously the SMA overlay had much lower density of distress than the AC overlay, indicating that the performance of the overlay mainly depended on upper layer.

Figure 7 shows that the geotextile had moved upward, indicating that the overlay suffered from the high shear deformation induced by the poor joint load transfer coefficient, and thus resulting in that the top of the overlay suffered high tensile stress, leading to Top-Down cracking.

Fig. 7. The geotextile had moved upward

7 Conclusions

The conclusions are summarizes below:

(1) All the reflective cracks are Top-Down Cracking. The mechanism for this is the poor of the load transfer coefficient of joints.
(2) The SMA asphalt mixture top layer performed much better than AC top layer, and thus should be adopted as the top layer of the asphalt overlay to delay the reflective cracking.
(3) Prior to overlaying the asphalt layer the load transfer efficiency of joints of existing concrete pavements should be improved to delay the reflective cracking.

Acknowledgements. The work was supported by the Guangxi Transportation Science and Technology Fund Project of China (Grant Nos. GNGKJGC2014-010), the Nanning City Science and Technology Fund Project of Guangxi, China (Grant Nos. 20171002-2).

References

Harmelink, D., Shuler, S., Aschenbrener, T.: Top-down cracking in asphalt pavements: causes, effects, and cures. J. Transp. Eng. **134**(1), 1–6 (2008)

Du, R., Lin, Y., et al.: Evaluation of composite pavements in Liuzhou-Nanning expressway in Guangxi province-Technical Report (2016) (in Chinese)

Chen, F.-F., Huang, X., Yue, X.: Analysis of factors to reflection crack of AC layer in CRC-AC compound pavement. J. Highw Transp. Res. Dev. **24**(2), 48–51 (2007)

An, J., Nam, B., Kim, J.: Numerical analysis of reflective cracking in an asphalt concrete overlay over a flexible pavement design, analysis, and asphalt material characterization for road and airfield pavements. GSP **246**, 67–74 (2014)

Zeng, M., Wu, Z., Zhong, M., Luo, D.: Finite element analysis of reflection crack in AC + CRC. J. Beijing Univ. Technol. **39**(10), 1516–1520 (2013). (in Chinese)

Li, S., Liu, Z., Li, Y.: Crack propagation in AC layer of continuously reinforced concrete composite asphalt pavement. J. Highw. Transp. Res. Dev. **28**(12), 2–9 (2011). (in Chinese)

Optimization of Gravel Road Blading

Wynand J. vd M. Steyn(✉)

HOD, Department of Civil Engineering, University of Pretoria,
Pretoria, South Africa
wynand.styen@up.ac.za

Abstract. Blading maintenance of gravel roads is important to ensure that the functional performance of the road is optimal and that vehicle operating costs are minimized. In order to optimize the blading cycles a simple, low-cost system is required to monitor the functional performance of the road and inform the blading need. A method is described where functional performance monitoring of gravel roads can be done at a regular frequency at no additional cost to the road owner. The effect of blading maintenance decisions taken based on the use of this system on a small gravel road network is illustrated. The optimization lead to less blading requirements as only the sections where functional performance was inadequate was maintained, and in the long term the condition of the road improved through regular appropriate blading.

Keywords: Gravel road blading · Functional performance · Optimization

1 Introduction

Roads (similar to all infrastructure) deteriorate with time and use and is thus in constant need of maintenance to ensure optimal functional condition (Jordaan 2013). Blading maintenance of gravel roads is one of the engineering options to ensure that the functional performance of the road is optimal and that vehicle operating costs are minimized. In order to optimize the blading cycles a system is required to monitor the functional performance of the road and inform the blading need. This system should preferably be low cost and simple to operate as the budgets available for such maintenance is typically low, and the condition needs to be monitored frequency due to the relatively quick deterioration in the functional performance of these roads through factors such as corrugation development.

A method is described where the monitoring of functional performance of gravel roads is conducted at a regular frequency and no additional cost to the road owner, using an innovative Response Type Road Roughness Measurement System (RTRRMS). The effect of blading maintenance decisions taken based on the use of this system on a small gravel road network is illustrated. The optimization lead to less blading requirements as only the sections where functional performance is inadequate are maintained, and in the long term the condition of the road improved through regular appropriate blading.

K. Zhang et al. (eds.), *Testing and Characterization of Asphalt Materials and Pavement Structures*, Sustainable Civil Infrastructures,
https://doi.org/10.1007/978-3-319-95789-0_15

2 Background to Road Maintenance and Roughness

2.1 Road Maintenance

Road maintenance is the process where the condition of a road needs to be monitored at regular intervals, and this condition data are used to make decisions regarding the required maintenance. The condition monitoring can include aspects such as the visual condition, functional condition and structural condition of the road (SADC 2003; SAPEM 2014). One of the functional condition aspects is the riding quality of the road. The level of riding quality (or road roughness) is directly linked to road safety, user perception and road user costs on the road (Heggie 1995; Sayers and Karamihas 1998).

In order to manage the maintenance of roads, minimum standards are set for the condition level of the various condition parameters on a road. In South Africa, the current guideline for riding quality of gravel roads indicates a minimum riding quality of 7 m/km on such roads (Henderson and Van Zyl 2017).

2.2 Riding Quality

The riding quality of a road can be monitored using two main types of monitoring (Jooste 1997) which are Response Type Road Roughness Measurements (RTRRM) and Profilometric Type (PT) measurements. Each of these approaches have distinct advantages and limitations, with RTRRM Systems (RTRRMS) using the calibrated response of a vehicle travelling over the road to indicate the roughness level while profilometric methods uses a laser-based measurement of the profile of the road surface that is mathematically analyzed using the Quarter-Car model to calculate the roughness. Both systems provide a roughness level in terms of the International Roughness Index (IRI). Which provides a mathematical indication of the physical response of a standardized vehicle to the road profile. The typical IRI scale runs from 0 m/km (perfect surface) to 20 m/km (impassable).

In general, the profilometric approach is more modern and sophisticated, and provides more consistent data. However, the approach requires significantly more expensive equipment and in-depth understanding and monitoring of the measured data, and the system can typically not be used on unpaved or gravel roads due to the dust generated during traveling on the road. The RTRRMS requires a calibrated vehicle that travels at a constant speed with a constant mass to ensure that the response of the vehicle to the road condition is not affected by external factors (Jooste 1997). Newer telematics-based RTRRMS uses algorithms that incorporate a range of speeds and vehicle types in the determination of the riding quality (Wessels and Steyn 2016).

3 Experiment and Methodology

This paper describes a process that was developed for the management of gravel road condition for a major agricultural business in the northern part of South Africa (Steyn and Coetzer 2015). The process required regular monitoring of the road condition as the deterioration of the road riding quality has a direct influence on the quality of the agricultural produce transported over the road system (Pretorius and Steyn 2016; Pretorius 2017).

The measurement system consisted of measuring the vertical accelerations on a Light Delivery Vehicle (LDV) on specific gravel routes before and after grading the route. Data were collected at 15 intervals over a period of 7 months. During this period the road was graded three times. The data analysis focuses on the comparison between the vertical acceleration data as measured using the LDV. All measurements were conducted traveling in the same direction on the route. Vehicle speed was kept constant by monitoring GPS-based speed and keeping the speed to 40 ± 5 km/h during any run.

An accelerometers was rigidly installed on the tow-bar of the LDV (Fig. 1), and the route was monitored using a calibrated GPS. The accelerometer data were collected at a frequency of 50 Hz and the focus was on the vertical accelerations. Collected acceleration data were analyzed through a set of algorithms to determine the Coefficient of Variation (CoV) of the vertical acceleration data that was then converted to Half-car Roughness Index (HRI) values using a set of calibration equations that were developed for the specific vehicle and vehicle conditions. Riding quality data were averaged over 250 m sections as these are deemed the shortest practical section lengths for grader-based road maintenance. The final data consisted of the locations and calculated riding quality over the length of the road at the selected intervals.

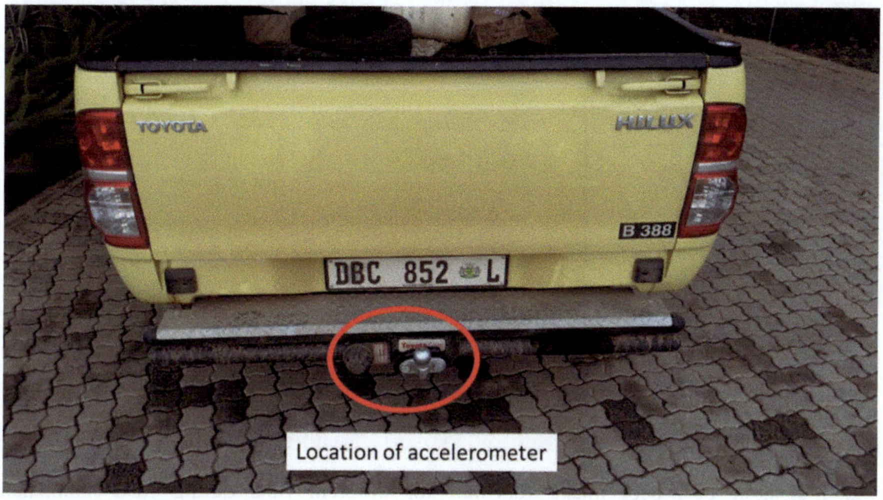

Fig. 1. Accelerometer rigidly attached to LDV tow bar

4 Data Analysis

In the data analysis attention was focused on the repeatability between measurement runs (vehicle location), and the measured acceleration response of the vehicles.

4.1 Vehicle Location

Vehicles traveling on a gravel route typically wander (i.e. travel on different wheel paths across the width of the road) across the road depending on the location of the smoothest ride on the road. A driver will typically try to select a route on the gravel road that prevents interaction with anomalies such as protruding rocks and corrugations. The actual routes followed by the LDV was analyzed by comparing the GPS-based location data over the length of the route. The typical accuracy of the GPS used was ± 4 m.

Analysis of the scatter in wheel tracks followed for the various runs on the road (shown for a unique short section in Fig. 2) indicates relative close spacing between the runs, with analysis of these wander data indicating that wander ranged between 1.9 and 6.7 m with an average of 4.0 m. If it is assumed that the average is due to GPS location issues and not physical wandering of the LDV, then the width of the wander ranges between an average of 0.0 and 2.4 m and thus the tires can be running on a totally different wheel track for the different measurements if care is not taken in the measurements. An experienced driver was appointed to conduct the measurements and this brought the wander to less than 0.5 m between runs.

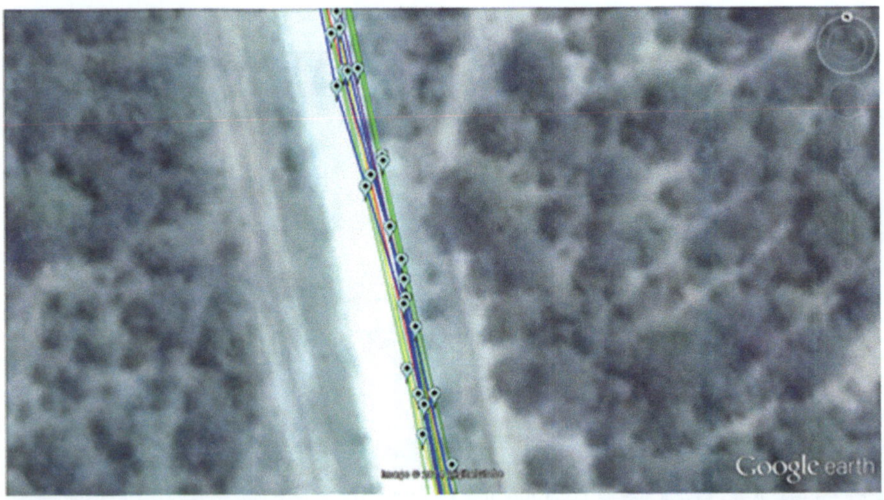

Fig. 2. Indication of wheel tracks followed for different runs during road maintenance measurements

4.2 Riding Quality—Acceleration Response

Analysis of all the vertical accelerations measured on the route at the intervals indicated generates the data in Fig. 3. It can be seen from the data that the Coefficient of Variation (CoV) of vertical acceleration (indicative of HRI) increases after grading the road (with use of the road by a range of vehicles), and then improves again after the next grading. In one instance a heavy rainfall caused relatively quick deterioration of the riding quality due to use of the road under very wet conditions. After ultimate rehabilitation of the road (which included major improvement in materials and construction options) the road condition remained good for a number of runs. Some of the variation in the data is attributable to vehicle wandering.

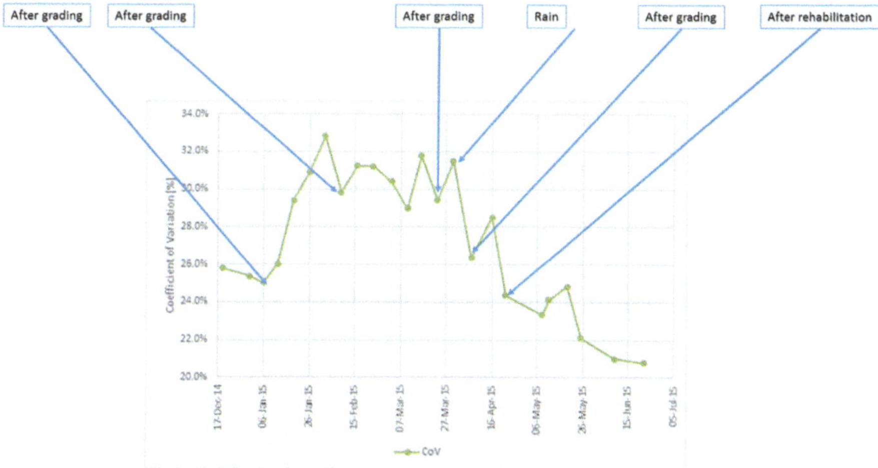

Fig. 3. Typical vertical acceleration data

4.3 Practical Application

In order for the data in the road maintenance analysis to be of practical value, a management value needs to be defined that can be used to monitor the change in the condition of the route until a certain trigger value is reached, at which stage grading of the route is recommended. In this regard the terminal riding quality value of 7 m/km (Henderson and Van Zyl 2017) is used in the current application of the process for the road owner. Regular measurement are conducted on the selected roads and a grading recommendation developed (Fig. 4) where only the sections with riding quality values more than the trigger value of 7 m/km are recommended to be maintained.

The grading recommendation consists of a short report highlighting the sections of the road that has higher than terminal riding quality and are thus in need of grading. Application of this process has saved the road owner between 40 and 60% of grading costs, as only sections that are worse than minimum riding quality are actually graded. A current process is focusing on the translation of the riding quality changes to the expected damage to transported tomatoes based on a set of relationships developed between the riding quality of the road and the effect on shelf-life of the tomatoes (Pretorius 2017). This provides a direct management tool to the agricultural company in terms that are applicable to the specific agricultural environment.

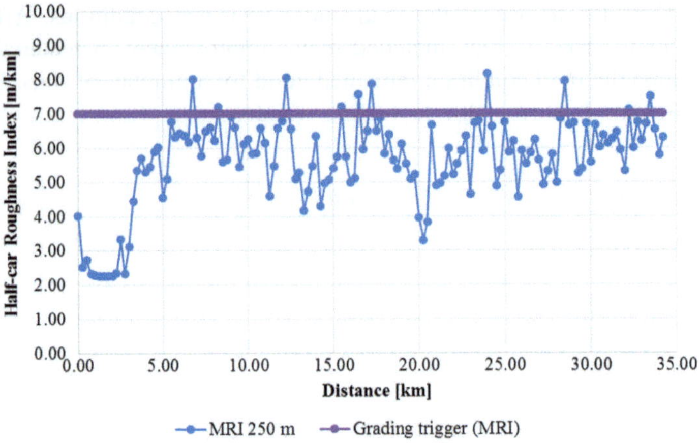

Fig. 4. Typical grading maintenance report

5 Conclusions

Based on the data collected in the project and discussed in this paper, the following conclusions are drawn:

- It is possible to measure the riding quality (and change in riding quality) of a gravel road consistently using a calibrated RTRRMS;
- The measured riding quality data can be used to monitor the effect of maintenance actions on the gravel road, and
- The measured riding quality data can be converted to a management tool for the road user that provides a direct understanding of the effect of the road condition on the economic operations on the road.

Acknowledgements. The support of the Tomato Producers Organization, ZZ2 and the University of Pretoria (financial support) are acknowledged. This work is partly based on the research supported in part by the National Research Foundation of South Africa. The Grantholder acknowledges that opinions, findings and conclusions or recommendations expressed in any publication generated by the NRF supported research are that of the authors, and that the NRF accepts no liability whatsoever in this regard.

References

Heggie, I.G.: Managing and Financing of Roads, an Agenda for Reform, p. 275. Washington D. C, World Bank Technical Paper number (1995)

Henderson, M., Van Zyl, G.: Management of unpaved roads: Developing a strategy and refining models. In: 36th Southern African Transport Conference (SATC), Pretoria, South Africa (2017)

Jooste, F.J.: Guidelines for network level measurement of road roughness. Version 1.1. South African Committee of Transport Officials (COTO), Pretoria, South Africa (1997)

Jordaan, G.: Optimisation of flexible road pavement rehabilitation investigations and design. LULU (2013)

Pretorius, C.J., Steyn, W.J.vdM.: The influence of road condition on the shelve-life of tomatoes. In: 4th Chinese European Workshop on Functional Pavement Design, Delft, Netherlands, 29 June–1 July 2016

Pretorius, C.J.: The influence of road condition on the shelf life of tomatoes. M.E. dissertation, University of Pretoria, Pretoria, South Africa (2017)

SADC: Guideline Low Volume Sealed Roads. Southern African Development Community (SADC), Gaborone, Botswana (2003). ISBN 99912-0-456-3

SAPEM: South African Pavement Engineering Manual, 2nd edn. South African National Roads Agency Soc Ltd., Pretoria, South Africa (2014). ISBN 978-1-920611-00-2

Sayers, M.W., Karamihas, S.M.: The Little Book of Profiling. University of Michigan, USA (1998)

Steyn, W.J.vdM., Coetzer, S.: ZZ2 Houtbosdorp bakkie grading measurements. University of Pretoria. UP/WJVDMS/02/2015, Pretoria, South Africa (2015)

Wessels, I., Steyn, W.J.vdM.: Real-time monitoring of the extended road network by utilising telematics technology. In: 4th Chinese European Workshop on Functional Pavement Design, Delft, Netherlands, 29 June–1 July 2016

Author Index

A
Abdelaziz, Amal, 90
Alam, Shahriar, 44
Ali, A., 1
Ali, Syed Ashik, 21
Assaf, Gabriel J., 136

B
Bors, John, 32

C
Chen, Can, 32
Chen, Xiangyang, 79

D
Dasek, Ondrej, 55
Du, Rongyao, 151

E
Eisenhut, Wolfgang O., 32

G
Ghabchi, Rouzbeh, 21
Guha, Aioub H., 136

H
Ho, Chun-Hsing, 90
Hossain, Zahid, 44
Huang, Yishuo, 12

I
Ingram, Alex, 32

J
Jaskula, Piotr, 104

L
Lau, Kreisler, 32
Liang, Tongxin, 128
Lin, Yuan-Zhih, 12
Liu, Shitao, 128

M
Mehta, Y., 1

P
Purdy, C., 1

R
Ranadive, M. S., 118
Rani, Shivani, 21
Rizvi, Hashim, 1
Rys, Dawid, 104

S
Shi, Jinjin, 69, 128
Spacek, Petr, 55
Su, Yu-Min, 12
Suda, Jan, 55

T
Tapase, Anand, 118
vd M. Steyn, Wynand J., 163

W
Weis, D., 1
Wu, Shenming, 12
Wu, Yingbiao, 69

X
Xiao, Qingyi, 69, 79

K. Zhang et al. (eds.), *Testing and Characterization of Asphalt Materials and Pavement Structures*, Sustainable Civil Infrastructures,
https://doi.org/10.1007/978-3-319-95789-0

Y
Yang, Yun, 79
Yu, Hongling, 128

Z
Zak, Josef, 55
Zaman, Musharraf, 21
Zhang, Peiliang, 69
Zhang, Ying, 128
Zhao, Jinguo, 79

MIX
Papier aus verantwortungsvollen Quellen
Paper from responsible sources
FSC® C105338

If you have any concerns about our products,
you can contact us on
ProductSafety@springernature.com

In case Publisher is established outside the EU,
the EU authorized representative is:
Springer Nature Customer Service Center GmbH
Europaplatz 3, 69115 Heidelberg, Germany

Printed by Libri Plureos GmbH
in Hamburg, Germany